QOS-ENABLED NETWORKS

WILEY SERIES IN COMMUNICATIONS NETWORKING & DISTRIBUTED SYSTEMS

Edited by
JOE SVENTEK,
DAVID HUTCHISON
SERGE FDIDA

Software Defined Mobile Networks (SDMN):
Beyond LTE Network Architecture
Madhusanka Liyanage (Editor),
Andrei Gurtov (Editor), *Mika Ylianttila* (Editor)

Publish / Subscribe Systems: Design and Principles
Sasu Tarkoma

Mobility Models for Next Generation Wireless
Networks: Ad Hoc, Vehicular and Mesh Networks
Paolo Santi

QOS-Enabled Networks: Tools and Foundations
Miguel Barreiros, Peter Lundqvist

MPLS-Enabled Applications: Emerging
Developments and New Technologies, 3rd Edition
Ina Minei, Julian Lucek

Personal Networks: Wireless Networking for
Personal Devices
*Martin Jacobsson, Ignas Niemegeers, Sonia
Heemstra de Groot*

Network Mergers and Migrations: Junos
Design and Implementation
*Gonzalo Gómez Herrero, Jan Antón
Bernal van der Ven*

Core and Metro Networks
Alexandros Stavdas

6LoWPAN: The Wireless Embedded Internet
Zach Shelby, Carsten Bormann

Mobile Peer to Peer (P2P): A Tutorial Guide
Frank H. P. Fitzek (Editor), *Hassan Charaf* (Editor)

Inter-Asterisk Exchange (IAX): Deployment
Scenarios in SIP-Enabled Networks
Mohamed Boucadair

MPLS-Enabled Applications: Emerging
Developments and New Technologies, 2nd Edition
Ina Minei, Julian Lucek

Host Identity Protocol (HIP): Towards the Secure
Mobile Internet
Andrei Gurtov

Service Automation and Dynamic Provisioning
Techniques in IP / MPLS Environments
*Christian Jacquenet, Gilles Bourdon,
Mohamed Boucadair*

Towards 4G Technologies: Services
with Initiative
Hendrik Berndt (Editor)

Fast and Efficient Context-Aware Services
*Danny Raz, Arto Tapani Juhola,
Joan Serrat-Fernandez, Alex Galis*

The Competitive Internet Service Provider:
Network Architecture, Interconnection,
Traffic Engineering and Network Design
Oliver M. Heckmann

Network Congestion Control: Managing
Internet Traffic
Michael Welzl

Service Provision: Technologies for Next
Generation Communications
Kenneth J. Turner (Editor),
Evan H. Magill (Editor),
David J. Marples (Editor)

Grid Computing: Making the Global
Infrastructure a Reality
Fran Berman (Editor), *Geoffrey Fox* (Editor),
Anthony J. G. Hey (Editor)

Web-Based Management of IP Networks
and Systems
Jean-Philippe Martin-Flatin

Security for Ubiquitous Computing
Frank Stajano

Secure Communication: Applications
and Management
Roger Sutton

Voice over Packet Networks
David J. Wright

Java in Telecommunications:
Solutions for Next Generation Networks
Thomas C. Jepsen (Editor),
*Farooq Anjum, Ravi Raj Bhat, Ravi Jain,
Anirban Sharma, Douglas Tait*

QOS-ENABLED NETWORKS
TOOLS AND FOUNDATIONS

SECOND EDITION

Miguel Barreiros
Juniper Networks, Portugal

Peter Lundqvist
Arista Networks, Sweden

This edition first published 2016
© 2016 John Wiley & Sons, Ltd

First Edition published in 2011

Registered Office
John Wiley & Sons, Ltd, The Atrium, Southern Gate, Chichester, West Sussex, PO19 8SQ,
United Kingdom

For details of our global editorial offices, for customer services and for information about how
to apply for permission to reuse the copyright material in this book please see our website at
www.wiley.com.

Library of Congress Cataloging-in-Publication data applied for

ISBN: 9781119109105

A catalogue record for this book is available from the British Library.

Set in 11/14pt Times by SPi Global, Pondicherry, India

1   2016

Contents

About the Authors

Miguel Barreiros is the Data Center Practice Lead at Juniper Networks responsible for the EMEA region. Previously he was a Senior Solutions Consultant focused on both data centers and IP/MPLS networks. Since he joined Juniper Networks in 2006, he has focused on the creation and development of solutions and has been involved in projects that span all stages of building and expanding networks, from design and testing to implementation and ongoing maintenance. He began his networking career in 2000, when as a hobby he was a network administrator for a British multiplayer gaming website that hosted network servers for various video games. Miguel has a B.Sc. degree in Electronics and Computer Engineering from Instituto Superior Técnico. He holds Juniper Networks Certificate Internet Expert (JNCIE) 193 and is a Juniper Networks Certified Instructor.

Peter Lundqvist is a System Engineer in Arista Networks since 2014, focusing on Datacenter solutions. Previously Peter worked at Juniper Networks in various roles including Juniper Networks Strategic Alliance group with a focus on packet-based mobile networks. Earlier Peter was a member of the Juniper Networks Beta Engineering team, which is responsible for early field testing of new hardware products and software versions. Peter's focus was on routing and control plane protocols. Before joining Juniper in 2000, Peter worked at Cisco Systems as a Consulting Engineer and at Ericsson. Peter has a degree in Systems and Information Technology from Mittuniversitetet (Mid Sweden University). He holds Juniper Networks Certificate Internet Expert (JNCIE) 48 and Cisco Certified Internetwork Expert (CCIE) 3787.

Foreword

Network consolidation has been with us since the 1990s, driven by the simple requirement to reduce the costs of business communication. For IT, it is a matter of controlling CapEx and OpEx. For service providers, it is a matter of offering multiservice solutions at a competitive cost. (Remember when "triple play" was the buzzword of the day?) Consolidation has been so successful that you seldom encounter an organization these days that runs separate data and telephony networks. Voice and video over IP is proven, reliable, and cheap. And modern service providers—whether they got their start as a telephony, cable, long distance, or Internet provider—now run all of their services over an IP core.

Treating all communications as data, and sending it all over a shared IP infrastructure—or series of IP infrastructures—has also revolutionized our modern lives from smart phones to shopping to entertainment to travel. For myself, one of the most interesting impacts of technology has been how different my teenagers' social lives are from my own when I was a teenager. Their activities are more spontaneous, their social groups are larger, and always-available communications make their activities safer.

And consolidation is still evolving. These days the excitement is around virtualization, improving the utilization of our existing communications resources.

From the beginning, one of the biggest challenges of consolidating all communications onto an IP infrastructure stems from the fact that not all data is equal. As users we expect a certain Quality of Experience (QOE) related to the service we're using. So QOE for voice is different than QOE for videoconferencing, both of which are different from high-definition entertainment. Each kind of data stream requires different treatment within the network to meet users' QOE expectations, and that's where Quality of Service (QOS) comes in.

QOS has been around as long as IP has. The IP packet header has a Type of Service (TOS) field for differentiating services, and over the years that field has evolved into the more sophisticated Differentiated Services Code Point (DSCP) field to better fit modern QOS classification strategies. And from the beginning it was understood that although IP provides connectionless best-effort delivery, some applications need reliable, sequenced, connection-oriented delivery. Hence TCP, which "fakes" the behavior of a wired-up point-to-point connection over IP.

QOS is really all about managing limited network resources. You don't get extra bandwidth or faster delivery; you just get to decide what data gets first dibs at the available resources. High-Def video requires very prompt delivery. A web page can wait a bit longer, and e-mail can wait much longer still. Over the years, QOS technologies and strategies have become more and more sophisticated to deal with the diversity of applications using the network. Routers and switches have better and better queues and queuing algorithms, better ingress control mechanisms, and better queue servicing mechanisms. And the advent of Software-Defined Networking (SDN) introduces some new and interesting ways of improving QOE.

All of this growing sophistication brings with it growing complexity for network architects and engineers. There are a lot of choices and a lot of knobs, and if you don't have the understanding to make the right choices and set the right knobs, you can do some serious damage to the overall quality of the network. Or at the least, you can fail to utilize your network's capabilities as well as you should.

That's where this book comes in. My longtime friends Miguel Barreiros and Peter Lundqvist have deep experience designing modern QOS strategies, and they share that experience in this book, from modern QOS building blocks to applied case studies. They'll equip you well for designing the best QOS approach for your own network.

Jeff Doyle

Preface

Five years have elapsed between the original publishing of this book and this second edition, and it is unquestionably interesting to analyze what has changed. The original baseline was that Quality of Service, or QOS, was in the spotlight. Five years have elapsed and QOS prominence has just kept on growing. It has entered in new realms like the Data Center and also spread into new devices. It is no longer just switches and routers—now even servers have at their disposal a complete QOS toolkit to deal, for example, with supporting multiple virtual machines.

This book's focus remains in the roots and foundations of the QOS realm. Knowledge of the foundations of QOS is the key to understanding what benefits it offers and what can be built on top of it. This knowledge will help the reader engage in both the conceptual and actual tasks of designing or implementing QOS systems, thinking in terms of the concepts, rather than thinking of QOS simply as a series of commands that should be pasted into the configuration of the devices. It will also help the reader to troubleshoot a QOS network, to decide whether the undesired results being seen are a result of misconfigured tools that require some fine-tuning or the wrong tools. As Galileo Galilei once said, "Doubt is the father of all invention."

A particular attention is also dedicated to special traffic types and networks, and three case studies are provided where the authors share their experience in terms of practical deployments of QOS.

Although the authors work for two specific vendors, this book is completely vendor agnostic, and we have shied away from showing any CLI output or discussing hardware-specific implementations.

History of This Project

The idea behind this book started to take shape in 2007, when Miguel engaged with British Telecom (BT) in several workshops about QOS. Several other workshops and training initiatives followed, and the material presented matured and stabilized over time. In July 2009, Miguel and Peter, who had also developed various QOS workshop and training guides, joined together to work on this project which led to the creation of the first edition.

In December 2014, both authors agreed that the book needed a revamp to cover the new challenges posed in the Data Center realm, which originated this second edition.

Who Should Read This Book?

The target audience for this book are network professionals from both the enterprise and the service provider space who deal with networks in which QOS is present or in which a QOS deployment is planned. Very little knowledge of other areas of networking is necessary to benefit from this book, because as the reader will soon realize, QOS is indeed a world of its own.

Structure of the Book

This book is split into three different parts following the Julius Caesar approach ("Gallia est omnis divisa in partes tres"):

Part One provides a high-level overview of the QOS tools. It also discusses the challenges within the QOS realm and certain types of special traffic and networks.

Part Two dives more deeply into the internal mechanisms of the important QOS tools. It is here that we analyze the stars of the QOS realm.

Part Three glues back together all the earlier material in the book. We present three case studies consisting of end-to-end deployments: the first focused on VPLS, the second focused on Data Center, and the third one focused on the mobile space.

Have fun.

Miguel Barreiros, *Sintra, Portugal*
Peter Lundqvist, *Tyresö, Sweden*
April 30, 2015

Acknowledgments

Several colleagues of ours have helped us during this project. However, three were absolutely key to the creation of this book:

Patrick Ames led this project in all its nontechnical aspects, allowing Miguel and Peter to focus on writing.

Aviva Garrett played the key role of the editorial review of the entire book and also guided Miguel and Peter in how to improve the book's organization and contents.

Steven Wong (Juniper JTAC) provided technical reviews for much of the book while sharing his immense knowledge of QOS.

The authors would also like to gratefully thank the following people: Antoine Sibout, Bobby Vandalore, Guy Davies, Jeff Doyle, John Peach, Jos Bazelmans, Kannan Kothandaraman, Margarida Correia, and Pedro Marques.

Finally, the authors would like to express their personal thanks:

Miguel: I would like to dedicate this book to Maria Eugénia Barreiros and to my grandparents José Silva and Dores Vilela.

Peter: My work on this book was possible only with the understanding and patience of the most dear ones in my life, my wife Lena and my great kids Ida and Oskar.

Abbreviations

2G	Second Generation
3GPP	Third-Generation Partnership Project
ACK	Acknowledgment
AF	Assured-forwarding
APN	Access Point Name
AUC	Authentication Center
BA	behavior aggregate
BE	best-effort
BHT	Busy Hour Traffic
Bps	bits per second
BSC	Base Station Controller
BSR	Broadband Service Router
BTS	Base Transceiver Station
BU	business
CDMA	Code Division Multiple Access
CEIR	Central Equipment Identity Register
CIR	Committed Information Rate
CLI	Command Line Interface
CNTR	control traffic
CoS	Class of Service
CT	class type
CWND	congestion window
DA	data
DF	Don't Fragment
DHCP	Dynamic Host Configuration Protocol
DiffServ	Differentiated Services

DNS	Domain Name System
DRR	Deficit Round Robin
DSCP	Differentiated Services Code Point
DSL	Digital Subscriber Line
DSLAM	Digital Subscriber Line Access Multiplexer
DWRR	Deficit Weighted Round Robin
EBGP	External Border Gateway Protocol
EF	Expedited-forwarding
EIR	Equipment Identity Register
EPC	Evolved Packet Core
ERO	Explicit Routing Object
eUTRAN	evolved UMTS Terrestrial Radio Access Network
FIFO	First in, first out
FQ	Fair queuing
GBR	Guaranteed Bit Rate
GGSN	Gateway GPRS Support Node
GPRS	General Packet Radio Service
GPS	Generic Processor Sharing
GSM	Global System for Mobile Communications
GTP	GPRS Tunneling Protocol
HLR	Home Location Register
ICMP	Internet Control Message Protocol
IMEI	International Mobile Equipment Identity
IMS	IP Multimedia System
IMSI	International Mobile Subscriber Identity
IntServ	Integrated Services
L2	Layer 2
L3	Layer 3
LBE	lower than that for best-effort
LFI	Link Fragmentation and Interleaving
LSP	label-switched path
LTE	Long-Term Evolution
MAD	dynamic memory allocation
ME	Mobile Equipment
MED	multi-exit discriminator
MF	Multifield
MME	Mobility Management Entity
MPLS	Multiprotocol Label Switching
MPLS-TE	MPLS network with traffic engineering

MS	Mobile System
ms	milliseconds
MSC	Mobile Switching Center
MSS	Maximum Segment Size
MTU	Maximum Transmission Unit
NAS	Non-Access Stratum
NC	Network-control
P2P	point-to-point
PB-DWRR	Priority-based deficit weighted round robin
PCR	Program Clock Reference
PCRF	Policy and Charging Rules Function
PDN	Packet Data Networks
PDN-GW	Packet Data Network Gateway
PDP	Packet Data Protocol
PE	provider edge
PHB	per-hop behavior
PID	Packet ID
PIR	peak information rate
PLMN	Public LAN Mobile Network
PMTU	Path MTU
pps	packets per second
PQ	priority queuing
PSTN	Public Switched Telephone Network
Q0	queue zero
Q1	queue one
Q2	queue two
QCI	QOS Class Identifier
QOS	Quality of Service
RAN	Radio Access Networks
RED	Random Early Discard
RNC	Radio Network Controller
RSVP	Resource Reservation Protocol
RT	real time
RTCP RTP	Control Protocol
RTT	Round Trip Time
SACK	selective acknowledgment
SAE	System Architecture Evolution
SCP	Secure Shell Copy
SCTP	Stream Control Transmission Protocol

SDP   Session Description Protocol
SGSN   Serving GPRS Support Node
S-GW   Serving Gateway
SIM   Subscriber Identity Module
SIP   Session Initiation Protocol
SLA   service-level agreement
SSRC   Synchronization Source Identifier
STP   Spanning Tree Protocols
TCP   Transmission Control Protocol
TE   Traffic Engineering
TOS   Type of Service
TS   Transport Stream
UDP   USER Datagram Protocol
UE   User Equipment
UMTS   Universal Mobile Telecommunications System
UTP   Unshielded Twisted Pair
VLAN   Virtual LAN
VLR   Visitor Location Register
VoD   Video on Demand
VoIP   Voice over IP
VPLS   Virtual Private LAN Service
VPN   Virtual Private Network
WFQ   Weighted Fair Queuing
WRED   Weighted RED
WRR   Weighted Round Robin

Part I
The QOS Realm

1

The QOS World

Quality of Service (QOS) has always been in a world of its own, but as the technology has been refined and has evolved in recent years, QOS usage has increased to the point where it is now considered a necessary part of network design and operation. As with most technologies, large-scale deployments have led to the technology becoming more mature, and QOS is no exception.

The current trend in the networking world is convergence, abandoning the concept of several separate physical networks in which each one carries specific types of traffic, moving toward a single, common physical network infrastructure. This is old news for the Internet and other service providers, however, a novelty in other realms such as the Data Center. The major business driver associated with this trend is cost reduction: one network carrying traffic and delivering services that previously demanded several separate physical networks requires fewer resources to achieve the same goal.

One of the most striking examples is voice traffic, which was previously supported on circuit-switched networks and is now delivered on the "same common" packet-switched infrastructure. Also, in modern Data Centers the operation of a server writing into the hard drive, the disk, is done using a networking infrastructure that is shared with other traffic types.

The inherent drawback in having a common network is that the road is now the same for different traffic types, which poses the challenge of how to achieve a peaceful coexistence among them since they are all competing for the same network resources.

QOS-Enabled Networks: Tools and Foundations, Second Edition. Miguel Barreiros and Peter Lundqvist.
© 2016 John Wiley & Sons, Ltd. Published 2016 by John Wiley & Sons, Ltd.

Allowing fair and even competition by having no traffic differentiation does not work because different types of traffic have different requirements, just like an ambulance and a truck on the same road have different needs. There is always the temptation of simply making the road wider, that is, to deploy network resources in an over-provisioned manner following the logic that although the split of resources was not ideal, so many free resources would be available at all times that the problem would be minimized. However, this approach has some serious drawbacks. First, in certain networks, the traffic flows and patterns are not predictable making it impossible to know the required resources beforehand. Secondly, it works against the major business driver behind network convergence, which is cost reduction. And third, such over-provisioning needs to be done not only for the steady state but also to take into account possible network failure scenarios.

QOS does not widen the road. Rather, it allows the division of network resources in a nonequal manner, favoring some and shortchanging others instead of offering an even split of resources across all applications. A key point with QOS is that a nonequal split of resources implies that there cannot be "win–win" situations. For some to be favored, others must be penalized. Thus, the starting point in QOS design is always to first select who needs to be favored, and the choice of who gets penalized follows as an unavoidable consequence.

In today's networks, where it is common to find packet-oriented networks in which different types of traffic such as voice, video, business, and Internet share the same infrastructure and the same network resources, the role of QOS is to allow the application of different network behaviors to different traffic types.

Hence, for a specific traffic type, two factors must be considered, characterizing the behavior that the traffic requires from the network and determining which QOS tools can be set in motion to deliver that behavior.

1.1   Operation and Signaling

The QOS concept is somewhat hard to grasp at first because it is structurally different from the majority of other concepts found in the networking world. QOS is not a standalone service or product but rather a concept that supports the attributes of a network by spanning horizontally across it.

QOS can be split into two major components: local operation and resource signaling. Local operation is the application of QOS tools on a particular router (or a switch, a server, or any QOS-capable device).

Resource signaling can be defined as the tagging of packets in such a way that each node in the entire path can decide which QOS tools to apply in a consistent fashion to assure that packets receive the desired end-to-end QOS treatment from the network.

These two components are somewhat similar to the IP routing and forwarding concepts. Routing is a task performed jointly by all routers in the network. All routers exchange information among them and reach a consistent agreement in terms of the end-to-end path that packets follow. As for forwarding, each router performs the task individually and independently from the rest of the network using only local information.

Routing is comparatively more complex than forwarding, because it involves cooperation among all the routers in the network. However, routing does not need to work at wire speed. Forwarding is simpler. It is a task performed by a router individually and independently. However, it must operate at wire speed.

An analogy between routing and forwarding, and QOS resource signaling and local operation, can be drawn. QOS resource signaling is somewhat analogous to the routing concept. It involves all routers in the network but has no requirement to work at wire speed. QOS local operation is analogous to the forwarding concept. Like forwarding, QOS local operation is, in concept, simpler, and each router performs it independently and individually. Also, QOS local operation must operate at wire speed.

However, there is a major difference between QOS resource signaling and routing; there are no standardized specifications (such as those which exist for any routing protocol) regarding what is to be signaled, and as a result there is no standard answer for what should be coded on all network routers to achieve the desired end-to-end QOS behavior. The standards in the QOS world do not give us an exact "recipe" as they do for routing protocols.

1.2   Standards and Per-Hop Behavior

The two main standards in the IP realm that are relevant to QOS are the Integrated Services (IntServ) and the Differentiated Services (DiffServ). IntServ is described in RFC1633 [1] and DiffServ in RFC2475 [2].

IntServ was developed as a highly granular flow-based end-to-end resource reservation protocol, but because of its complexity, it was never commonly deployed. However, some of its concepts have transitioned to the MPLS world, namely, to the Resource Reservation Protocol (RSVP).

The DiffServ model was developed based on a class scheme, in which traffic is classified into classes of service rather than into flows as is done with IntServ. Another major difference is the absence of end-to-end signaling, because in the DiffServ model each router effectively works in a standalone fashion.

With DiffServ, a router differentiates between various types of traffic by applying a classification process. Once this differentiation is made, different QOS tools are applied to each specific traffic type to effect the desired behavior. However, the standalone model used by DiffServ reflects the fact that the classification process rules and their relation to which QOS tools are applied to which type of traffic are defined locally on each router. This fundamental QOS concept is called per-hop behavior (PHB).

With PHB, there is no signaling between neighbors or end to end, and the QOS behavior at each router is effectively defined by the local configuration on the router. This operation raises two obvious concerns. The first is how to achieve coherence in terms of the behavior applied to traffic that crosses multiple routers, and the second is how to propagate information among routers.

Coherence is achieved by assuring that the routers participating in the QOS network act as a team. This means that each one has a consistent configuration deployed which assures that as traffic crosses multiple routers, the classification process on each one produces the same match in terms of which different traffic types and which QOS tools are applied to the traffic.

Unfortunately, the PHB concept has its Achilles' heel. The end-to-end QOS behavior of the entire network can be compromised if a traffic flow crosses a number of routers and just one of them does not apply the same consistent QOS treatment, as illustrated in Figure 1.1.

In Figure 1.1, the desired behavior for the white packet is always to apply the PHB A. However, the middle router applies a PHB different from the desired one, breaking the desired consistency across the network in terms of the QOS treatment applied to the packet.

The word *consistent* has been used frequently throughout this chapter. However, the term should be viewed broadly, not through a microscopic perspective. Consistency does *not* mean that all routers should have identical configurations. Also, as we will see, the tools applied on a specific router vary according to a number of factors, for example, the router's position in the network topology.

The second challenge posed by the PHB concept is how to share information among routers because there is no signaling between neighbors or end to end. Focusing on a single packet that has left an upstream router and is arriving at the downstream router, the first task performed on that packet is

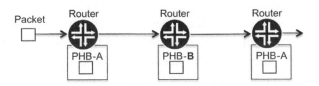

Figure 1.1   End-to-end consistency

classification. The result of this classification is a decision regarding which behavior to apply to that packet. For instance, if the upstream router wants to signal information to its neighbor regarding this specific packet, the only possible way to do so is to change the packet's contents by using the rewrite QOS tool, described in Chapter 2. Rewriting the packet's content causes the classification process on the downstream router to behave differently, as illustrated in Figure 1.2.

However, the classification process on the downstream router can simply ignore the contents of the packet, so the success of such a scheme always depends on the downstream router's consistency in terms of its classifier setup. A somewhat similar concept is the use of the multi-exit discriminator (MED) attribute in an External Border Gateway Protocol (EBGP) session. The success of influencing the return path that traffic takes depends on how the adjacent router deals with the MED attribute.

Although it does pose some challenges, the DiffServ/PHB model has proved to be highly popular. In fact, it is so heavily deployed that it has become the de facto standard in the QOS realm. The reasons for this are its flexibility, ease of implementation, and scalability, all the result of the lack of end-to-end signaling and the fact that traffic is classified into classes and not flows, which means that less state information needs to be maintained among the network routers. The trade-off, however, is the lack of end-to-end signaling, which raises the challenges described previously. But as the reader will see throughout this book, these issues pose no risk if handled correctly.

As an aside, in Multiprotocol Label Switching (MPLS) networks with Traffic Engineering (TE), it is possible to create logical paths called label-switched paths (LSPs) that function like tunnels across the network. Each tunnel has a certain amount of bandwidth reserved solely for it end to end, as illustrated in Figure 1.3.

What MPLS-TE changes in terms of PHB behavior is that traffic that is placed inside an LSP has a bandwidth assurance from the source to the destination. This means, then, that in terms of bandwidth, the resource competition is limited to traffic inside that LSP. Although an MPLS LSP can have a bandwidth reservation,

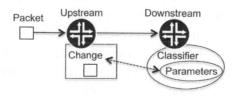

Figure 1.2   Signaling information between neighbors

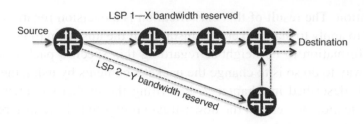

Figure 1.3   MPLS-TE bandwidth reservations

it still requires a gatekeeper mechanism at the ingress node to ensure that the amount of traffic inserted in the LSP does not exceed the reserved bandwidth amount.

Another difference is that MPLS-TE allows explicit specification of the exact path from the source to the destination that the traffic takes instead of having the forwarding decision made at every single hop. All other PHB concepts apply equally to QOS and MPLS.

MPLS is a topic on its own and is not discussed more in this book. For more information, refer to the further reading section at the end of this chapter.

1.3   Traffic Characterization

As we have stated, different traffic types require that the network behave differently toward them. So a key task is characterizing the behavioral requirements, for which there are three commonly used parameters: delay, jitter, and packet loss.

For an explanation of these three parameters, let's assume a very simplistic scenario, as illustrated in Figure 1.4. Figure 1.4 shows a source and an end user connected via a network. The source sends consecutively numbered packets 1 through 3 toward the end user. Packet 1 is transmitted by the source at time t_1 and received by the end user at the time r_1. The same logic applies for packets 2 and 3. A destination application is also present between the end user and the network, but for now its behavior is considered transparent and we will ignore it.

Delay (also commonly called latency) is defined as the time elapsed between the transmission of the packet by the source and the receipt of the same packet by the destination (in this example, the end user). In Figure 1.4, for packet 1, delay is the difference between the values r_1 and t_1, represented by the symbol Δ_1, and is usually measured in milliseconds.

Jitter represents the variation in delay between consecutive packets. Thus, if packet 1 takes Δ_1 to transit the network, while packet 2 takes Δ_2, then the jitter between packets 1 and 2 can be seen as the difference between Δ_1 and Δ_2 (also measured in milliseconds).

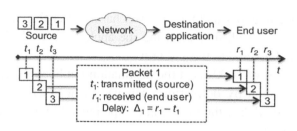

Figure 1.4   Delay, jitter, and packet loss across the network

The other parameter paramount to QOS traffic characterization is packet loss. This parameter represents how many packets are not received compared with the total number of packets transmitted and is usually measured as a percentage.

In terms of the sensitivity that traffic has to these three parameters, it is important to differentiate between real-time and nonreal-time traffic. There is also a third special traffic type, storage, but due to its uniqueness it is detailed in a dedicated section in Chapter 4. For real-time traffic, the main focus sensitivities are generally delay and jitter. So let's start with these two parameters, and we'll focus on packet loss a little later on.

Delay is important because real-time packets are relevant to the destination only if they are received within the time period in which they are expected. If that time period has expired, the packets become useless to the destination. Receiving them not only adds no value but also has a negative impact because although the packet is already useless, receiving it still demands processing cycles at the destination.

Jitter can also be very problematic because it interrupts the consistency of the delay of the packets arriving at destination. This interruption poses serious problems to the application receiving the traffic by forcing it to be constantly adapting to new delay values. Practical experience from voice-over-IP (VoIP) deployments shows that users migrating from a circuit-switched network can easily get used to a delay value even slightly higher than what they previously had as long as it is constant. However, the presence of significant jitter immediately generates user complaints. The bottom line is that when the delay value is always changing, users (and applications) cannot get used to it because it is not constant.

Although the previous descriptions are generally applicable to various types of real-time traffic, they should not all be placed under the same umbrella, because the exact set of requirements depends on the application itself. For example, if the application using real-time traffic is unidirectional, buffering can be used at the destination to reduce the presence of jitter.

Looking again at Figure 1.4, assume that the traffic sent by the source to the end user is a unidirectional video stream. Also assume that the destination application placed between the network and the end user has a buffer that enables it to store the packets being received, thus allowing the application to decide when those packets should be delivered to the end user.

Assuming a buffer of 1000 ms at the destination application (enough to store all three packets), then by delivering each packet at a separation of 300 ms, which is the average delay, the jitter value experienced by the end user is zero, as illustrated in Figure 1.5.

The drawback to this solution is that it introduces delay, because packets are stored inside the destination application for a certain amount of time and are not immediately delivered to the end user. So there is a trade-off between reducing jitter and introducing delay.

As for the packet loss parameter, a packet of a real-time stream is useful for the destination only if received within a certain time period, a requirement that tends to invalidate any packet retransmission mechanism by the source in case of packet loss. Hence, it is no surprise that the User Datagram Protocol (UDP), a connectionless protocol, is commonly used for the transmission of real-time streams.

Different real-time applications have different levels of sensitivity to packet loss. For example, video applications generally display minor glitches or blocking when low-level loss occurs, but large packet loss can cause total loss of the picture. Similarly, for voice applications, a low-level loss generally causes minor clicks with which the human ear is perfectly capable of dealing. However, large-scale loss can simply cause the call to drop. Finding where to draw the line between what is an acceptable packet loss and what is a catastrophic packet loss scenario is highly dependent on the application.

For nonreal-time traffic, generally speaking, the sensitivity to jitter and delay is obviously much lower, because there is not such a strong correspondence between when the packet is received and the time interval in which the packet is useful for the destination.

As for packet loss, a split can be made regarding whether the application uses a connection-oriented protocol, such as the Transmission Control Protocol (TCP), or a connectionless protocol, such as UDP, for transport at OSI Layer 4. In the first scenario (TCP), the transport layer protocol itself takes care of any

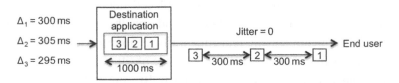

Figure 1.5   Jitter reduction by using a buffer at the destination application

necessary packet retransmissions, while in the second scenario (UDP), the session layer (or a layer higher in the OSI stack) must handle the packet loss.

Another scenario is the network being lossless, meaning that the network itself will assure that there will not be any packet loss, thus removing the need for the transport or higher layers having to worry about that. The concepts of lossless network, UDP, and TCP are detailed further in Chapter 4.

As a teaser for the following chapters, we stated earlier in this chapter that QOS allows implementation of an unfair resource-sharing scheme across different traffic types. In these unfair schemes, offering benefit to some implies impairing others. So, for example, if real-time traffic is more sensitive to delay and jitter, QOS can allow it to have privileged access to network resources in terms of less delay and less jitter. Of course, this is achieved at the expense of possibly introducing more delay and jitter in other traffic types, which can be acceptable if they have higher tolerances to delay and jitter.

1.4   A Router without QOS

A useful starting point is to analyze the effects of the absence of QOS, which acts to provide a perspective on what the end result is that we want to achieve by the change.

In the scenario of a router without QOS enabled, the order of traffic present at the ingress interface is identical to the order of traffic as it leaves the router via the egress interface, assuming that no packet loss occurs, as illustrated in Figure 1.6.

Figure 1.6 shows two types of traffic, white and black, each one with three packets numbered 1 through 3. If QOS is not enabled on the router, the output sequence of packets at the egress interface is the same as it was at the ingress.

One of the many things that QOS allows is for a router to change that output sequence of packets with great exactness. Suppose black packets correspond to sensitive traffic that should be prioritized at the expense of delaying white packets, a behavior illustrated in Figure 1.7.

Figure 1.6   Traffic flow across a router without QOS

Figure 1.7   Router with QOS enables packet prioritization

To achieve this result requires differentiation: the router must be able to differentiate between white and black packets so it can make a decision regarding the output sequence order. This differentiation is achieved by classifying traffic into different classes of service.

1.5 Conclusion

QOS is all about not being fair when dividing the network resources but rather selecting discriminately who is favored and who gets penalized. QOS does not make the road wider; it just decides who goes first and as a consequence who has to wait.

In terms of standards, the key is the PHB concept, in which each router acts independently from the rest of the network in terms of the behavior it applies to the traffic that crosses it. PHB obligates the routers to consistently apply the desired behavior to each traffic type that crosses it in spite of its independent decision making.

In terms of the parameters used to define the traffic requirements, the more common ones are delay, jitter, and packet loss. The tolerance to these parameters is highly dependent on whether the traffic is real time or not, because for real-time traffic, the time gap in which the packet is considered useful for the destination is typically much narrower. An interesting development detailed in Chapter 4 is storage traffic, since it has zero tolerance regarding packet loss or reordering.

The chapters that follow in this book present the tools and challenges to achieve such traffic differentiation inside the QOS realm.

References

[1] Braden, R., Clark, D. and Shenker, S. (1994) RFC 1633, Integrated Services in the Internet Architecture: An Overview, June 1994. https://tools.ietf.org/html/rfc1633 (accessed August 19, 2015).
[2] Blake, S., Black, D., Carlson, M., Davies, E., Wang, Z. and Weiss, W. (1998) RFC 2475, Architecture for Differentiated Services, December 1998. https://tools.ietf.org/html/rfc2475 (accessed August 19, 2015).

Further Reading

Minei, I. and Lucek, J. (2011) *MPLS-Enabled Applications*. New York: John Wiley & Sons, Inc.

2

The QOS Tools

As with any technology, as QOS matures and evolves, new developments always occur and features are always being added. However, before diving deeply into the specifics and all possible variations and evolutions of each QOS tool, the first step is to understand the basic structure of each tool. This chapter presents a high-level view of the fundamental QOS tools. The complexities and specifics of these tools are explained in Part Two of this book.

In a general sense, a QOS tool is simply a building block. It receives input, processes it, and produces specific output. The relationship between the input and output depends on the internal mechanisms of the QOS tools.

Examining each QOS building block separately, focusing on the results each one delivers, allows us to determine how to combine them to achieve the desired behavior.

To explain the QOS building blocks, this chapter focuses on a single router that is part of a network in which QOS is implemented. The router has an ingress interface on which traffic arrives at the router and an egress interface from which traffic departs the router. It is between these two interfaces that QOS tools are applied to the traffic. We are using a router just as an example; it could have been any other QOS-capable device like a switch or a server.

2.1 Classifiers and Classes of Service

Classifiers perform a task that is somewhat analogous to an "identification" process. They inspect the incoming traffic, identify it, and then decide to which class of service it belongs.

QOS-Enabled Networks: Tools and Foundations, Second Edition. Miguel Barreiros and Peter Lundqvist.
© 2016 John Wiley & Sons, Ltd. Published 2016 by John Wiley & Sons, Ltd.

The basis of QOS is traffic differentiation, and the concept of a class of service is fundamental to the operation of QOS. Assigning traffic to different classes of service provides the necessary differentiation. From the point of view of a router, the class of service assigned to a packet defines how the router behaves toward it. The concept of traffic differentiation is present in every QOS tool, and as a result, classes of service are present across the entire QOS design.

A classifier has one input, the incoming packet, and it has N possible outputs, where N is the number of possible classes of service into which the packet can be classified. By definition, a classified packet is a packet assigned to a class of service. The decision made by the classifier regarding which class of service a packet belongs to can be seen as a set of IF/THEN rules, where the IF statement specifies the match conditions and, if those conditions are met, the THEN statement specifies the class of service into which the packet is classified, as illustrated in Figure 2.1.

Several possibilities exist for what can be used in the match conditions of the IF statements. The classifier can simply look at the ingress interface on which the packet has arrived and assign all traffic received on that interface to a certain class of service, based on a specific rule. Or, as another example, the classifier can use information that was placed in the packet headers by previous routers specifically for the purpose of telling downstream neighbors about a previous classification decision. More details about the range of possibilities regarding what can be used in the match conditions of the IF statement are given in Chapter 5.

Coming back to the analogy of the routing and forwarding world, when an IP packet arrives at a router, a route lookup is performed to decide where the packet is destined. The classifier concept is similar: the classifier inspects the packet and makes a decision regarding the class of service to which it belongs.

The classifier tool can also be used in a chain fashion. In a scenario of N chain classifiers, the next-in-chain classifier receives packets that have already been classified, meaning that a class of service has already been assigned. This previous classification provides one extra input to be used in the match conditions of the IF/THEN set of rules.

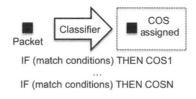

IF (match conditions) THEN COS1
...
IF (match conditions) THEN COSN

Figure 2.1   The classifier operation

Assuming for ease of understanding a scenario in which $N=2$, a packet crosses the first classifier and exits with a class of service assigned to it. Then it reaches the second classifier, which has three possible choices regarding the previously assigned class of service: (i) to accept it as is, (ii) to increase the granularity (e.g., by splitting "business" into "business in contract" and "business out of contract"), or (iii) to change it completely. (A complete change is a somewhat awkward decision and is usually reserved only when correcting a previous classification error.)

2.2   Metering and Coloring—CIR/PIR Model

In Chapter 1, we discussed parameters such as delay and jitter that allow differentiation between different traffic types. However, one further step can be taken in terms of granularity, which is to differentiate first between traffic types and then, within a traffic type, to differentiate even further. Returning to the road metaphor, this is akin to differentiating first between trucks and ambulances and then, for only ambulances, differentiating between whether the siren is on or off.

The metering tool provides this second level of granularity. It measures the traffic arrival rate and assigns colors to traffic according to that rate.

So looking at the classifier and metering tools together, the classifier assigns traffic to classes of service, which each have assigned resources, and the metering tool allows traffic inside a single class of service to be differentiated according to its arrival rate.

To achieve proper metering, we first present here a well-established model in the industry, the committed information rate (CIR)/peak information rate (PIR) model, which is inherited from the Frame Relay realm. However, it is first worth pointing out that this model is not part of any mandatory standard whose use is enforced. CIR/PIR is just a well-known model used in the networking world. One of the many references that can be found for it is RFC 2698 [1].

The foundations of this model are the definition of two traffic rates, CIR and PIR. CIR can be defined as the guaranteed rate, and PIR is the peak rate in terms of the maximum admissible traffic rate.

The CIR/PIR model has three traffic input rate intervals, where each has an associated color, as illustrated in Figure 2.2.

Traffic below the CIR is colored green (also commonly called *in-contract traffic*), traffic that falls in the interval between the CIR and PIR is colored yellow (also commonly called *out-of-contract traffic*), and traffic above the PIR is colored red.

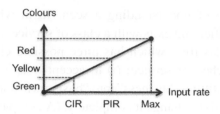

Figure 2.2   The CIR/PIR model

Figure 2.3   The metering tool

So following the CIR/PIR model, the metering tool has one input, the traffic arrival rate, and three possible outputs, green, yellow, or red, as illustrated in Figure 2.3.

Generally speaking, green traffic should have assured bandwidth across the network while yellow traffic should not, because it can be admitted into the network but it has no strict guarantee or assurance. Red traffic should, in principle, be discarded because it is above the peak rate.

Services that allow yellow traffic are popular and commonly deployed, following the logic that if bandwidth is available, it can be used under the assumption that this bandwidth usage does not impact other traffic types.

Commonly, metering is not a standalone tool. Rather, it is usually the subblock of the policing tool that is responsible for measuring the traffic input rate. The policing building block as a whole applies predefined actions to each different color.

2.3   The Policer Tool

The policer tool is responsible for ensuring that traffic conforms to a defined rate called the *bandwidth limit*. The output of the policer tool is the traffic that was present at input but that has been limited based on the bandwidth limit parameter, with excess traffic being discarded, as illustrated in Figure 2.4.

The behavior of discarding excess traffic is also called *hard policing*. However, several other actions are possible. For example, the excess traffic can be accepted but marked differently so it can be differentiated inside the router, a behavior commonly called *soft policing*.

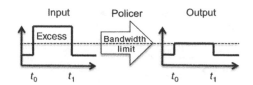

Figure 2.4   The policer tool

Figure 2.5   Combining the policer and the metering tools

The metering tool is commonly coupled with the policer as a way to increase the policer's granularity. In this scenario, the metering tool measures the traffic arrival rate and splits the traffic into the three-color scheme previously presented. The policer is responsible for applying actions to the traffic according to its color, as illustrated in Figure 2.5.

The policer can take one of several decisions regarding the traffic: (i) to transmit the traffic, (ii) to discard it, or (iii) to mark it differently and then transmit it.

Another capability associated with the policer tool is burst absorption. However, we leave this topic for Chapter 6, in which we introduce the use of the token bucket concept to implement the policer function.

In a scenario of linked meter and policer tools, the only difference is that the next step in the chain of meter and policer tool receives a packet that is already colored. This color provides an extra input parameter. The color previously assigned can be maintained, raised (e.g., from green to yellow), or lowered.

2.4   The Shaper Function

The shaper function causes a traffic flow to conform to a bandwidth value referred to as the *shaping rate*. Excess traffic beyond the shaping rate is stored inside the shaper and transmitted only when doing so does not violate the defined shaping rate. When traffic is graphed, you see that the format of the input traffic flow is shaped to conform to the defined shaping rate.

Let us illustrate this behavior using the example shown in Figure 2.6. Here, an input traffic flow exceeds the defined shaping rate (represented as a dotted line) in the time interval between t_0 and t_1, and packet X (represented as a gray ball) is the last packet in the input traffic flow between t_0 and t_1.

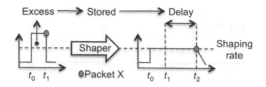

Figure 2.6   The shaper tool

Comparing the input and output traffic flow graphs, the most striking difference is the height of each. The height of the output traffic flow is lower, which is a consequence of complying with the shaping rate. The other difference between the two is the width. The output traffic flow is wider, because the traffic flows for a longer time, as a consequence of how the shaper deals with excess traffic.

Let us now focus on packet X, represented by the gray ball. At input, packet X is present at t_1. However, it is also part of excess traffic, so transmitting it at t_1 violates the shaping rate. Here, the shaper retains the excess traffic by storing it and transmits packet X only when doing so does not violate the shaping rate. The result is that packet X is transmitted at time t_2.

Effectively, the information present at the input and output is the same. The difference is its format, because the input traffic flow format was shaped to conform to the shaping rate parameter.

As Figure 2.6 also shows, the storage of excess traffic introduces delay into its transmission. For packet X, the delay introduced is the difference between t_2 and t_1.

As highlighted in Chapter 1, some traffic types are more sensitive to the delay parameter, so the fact that the shaping tool inherently introduces delay to sustain excess traffic can make it inappropriate for these traffic types.

Also, the shaper's ability to store excess traffic is not infinite. There is a maximum amount of excess traffic that can be retained inside the shaper before running out of resources. Once this point is reached, excess traffic is discarded, with the result that the amount of information present at the output is less than that present at input. We revisit this topic in Chapter 6 when discussing how to use the leaky bucket concept to implement the shaper tool.

As with many other QOS tools, the shaper can also be applied in a hierarchical fashion, something we explore in the VPLS case study in this book.

2.5   Comparing Policing and Shaping

A common point of confusion is the difference between policing and shaping. This confusion is usually increased by the fact that traffic leaving the policer conforms to the bandwidth limit value and traffic leaving the shaper conforms

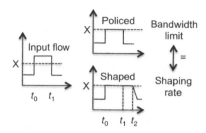

Figure 2.7   Comparing the policer and the shaper tools

to the shaping rate, so at a first glance the two may seem similar. However, they are quite different.

Consider Figure 2.7, in which the policer bandwidth limit and the shaping rate of the shaper are set to the same value, X, represented as a dotted line. The input traffic flow that crosses both tools is also equal, and between times t_0 and t_1 it goes above the dotted line.

The difference between the policed and the shaped flows is visible by comparing the two graphs. The main difference is how each tool deals with excess traffic. The policer discards it, while the shaper stores it, to be transmitted later, whenever doing so does not violate the defined shaping rate.

The result of these two different behaviors is that the policed flow contains less information than the input flow graph because excess traffic is discarded, whereas the shaped output flow contains the same information as the input traffic flow but in a different format, because excess traffic was delayed inside the shaper.

Another point of confusion is that the command-line interface (CLI) on most routers allows you to configure a burst size limit parameter inside the policer definition. However, doing this is not the same as being able to store excess traffic. All these differences are explained in Chapter 6 when we detail the token and leaky bucket concepts used to implement the policer and shaper tools, respectively.

The use of shaping to store excess traffic instead of dropping it has earned it the nickname of "TCP friendly," because for connection-oriented protocols such as TCP, it is usually better to delay than to drop packets, a concept that is detailed in Chapter 4 of this book.

2.6   Queue

A queue contains two subblocks: a buffer and a dropper. The buffer is where packets are stored while awaiting transmission. It works in a first in, first out (FIFO) fashion, which means that the order in which packets enter the buffer is

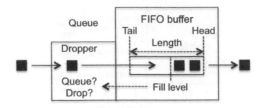

Figure 2.8 The queue operation

the same in which they leave. No overtaking occurs inside the buffer. The main parameter that defines the buffer is its length, which is how many packets it is able to store. If the buffer is full, all newly arrived packets are discarded.

For ease of understanding throughout this book, we always consider the situation where packets are placed in queues. However, this is not always the case. In certain queuing schemes, what is placed in the queue is in fact a notification cell, which represents the packet contents inside the router. We discuss this topic in detail in Chapter 7.

The decision whether a packet should be placed in the queue buffer or dropped is taken by the other queue block, the dropper. The dropper has one input, which is the queue fill level. Its output is a decision either to place the packet in the queue or to drop it. The dropper works in an adaptive control fashion, in which the decision to either place the packet in the queue or drop it is taken according to the queue fill level. So as the queue fills up or empties, the dropper behaves differently.

As a QOS tool, the queue has one input, which is the packets entering the queue, and one output, which is the packets leaving the queue, as illustrated in Figure 2.8.

When a packet enters a queue, it first crosses the dropper block, which, based on the queue fill level, makes a decision regarding whether the packet should be placed in the FIFO buffer or discarded. If the packet is placed in the buffer, it stays there until it is transmitted. The action of moving a packet out of the queue is the responsibility of the next QOS tool we discuss, the scheduler.

If the queue is full, the decision made by the dropper block is always to drop the packet. The most basic dropper is called the tail drop. Its basic behavior is that, when the queue fill level is less than 100%, all packets are accepted, and once the fill level reaches 100%, all packets are dropped. However, more complex dropper behaviors can be implemented that allow the dropping of packets before the queue buffer fill level reaches 100%. These behaviors are introduced in Chapter 3 and detailed in Chapter 8.

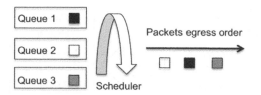

Figure 2.9   The scheduler operation

2.7   The Scheduler

The scheduler implements a multiplexing operation, placing N inputs into a single output. The inputs are the queues containing packets to be serviced, and the output is a single packet at a time leaving the scheduler. The scheduler services a queue by removing packets from it.

The decision regarding the order in which the multiple queues are serviced is the responsibility of the scheduler. Servicing possibilities range from fair and even treatment to treating some queues as being privileged in terms of having packets be processed at a faster rate or before other queues.

The scheduler operation is illustrated in Figure 2.9, which shows three queues, each containing a packet waiting for transmission. The scheduler's task is to decide the order in which queues are serviced. In Figure 2.9, queue 3 (Q3) is serviced first, then Q1, and finally Q2. This servicing order implies that the egress order of the packets is the gray packet first, then the black packet, and finally the white packet. Figure 2.9 provides just a simple scheduler example. In reality, there are multiple possible schemes regarding how the scheduler decides how to service the queues. These are discussed in Chapter 7.

In some ways, the scheduler operation is the star of the QOS realm. It is the tool by which the QOS principle of favoring some by penalizing others becomes crystal clear. We shall see in detail that if traffic is split into classes of service and each class is mapped into a specific queue, the scheduler provides a very clear and precise method to decide which traffic is favored by being effectively prioritized and transmitted more often.

2.8   The Rewrite Tool

The rewrite tool allows a packet's internal QOS marking (simply called its marking) to be changed. The exact location of the marking depends on the packet type, for example, whether it is an IP or MPLS packet, but for now simply assume that there is a marking inside the packet. We analyze the different packet types in Chapter 5.

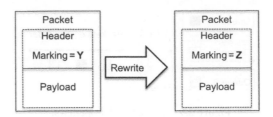

Figure 2.10 The rewrite operation

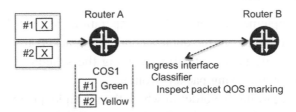

Figure 2.11 Information propagation

The rewrite tool has one input and one output. The input is a packet that arrives with a certain marking, and the output is the packet with the "new" marking, as illustrated in Figure 2.10.

Two major scenarios require application of the rewrite tool. In the first scenario, this tool is used to signal information to the next downstream router. The second scenario is when the packet marking is considered untrustworthy. In this case, the router at the edge of the network can rewrite the marking to ensure that the other network routers are not fooled by any untrustworthy marking in the packet. We now focus on the first scenario and leave the second one for Chapter 3, in which we discuss trust boundaries across a QOS network.

When we introduced the PHB concept in Chapter 1, we mentioned the lack of signaling between neighbors or end to end and raised the question of how to signal information between routers. Figure 2.11 illustrates this challenge with a practical example in which two packets, numbered 1 and 2, both with a marking of X, cross router A and are both classified into the class of service COS1. However, packet 1 is colored green and packet 2 yellow. At the other end, router B has a classifier that inspects the packets' QOS markings to decide to which class of service they belong.

If router A wants to signal to router B that the packets belong to the same class of service but they are colored differently, it has only one option because of the lack of signaling, which is to play with the packet-marking parameters used by router B's classification process. For example, if router A applies the

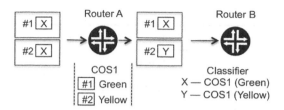

Figure 2.12   Rewrite tool applicability

Figure 2.13   Traffic flow across the router

rewrite tool to the second packet and changes its marking from X to Y, router B's classifier can differentiate between the two packets, as illustrated in Figure 2.12.

However, router A cannot control the classification rules that router B uses because of the lack of signaling. So the success of the rewriting scheme always depends on the router B classifier configuration being coherent so that the marking of X is translated into COS1 and color green and the marking of Y is also translated into COS1 but retains the color yellow. Consistency is crucial.

2.9   Example of Combining Tools

Now that we have presented each tool separately, we combine them all in this section in a simple walkthrough example to illustrate how they can be used together on a router. However, this should be seen as just an example that glues all the tools together. It is not a recommendation for how to combine QOS tools nor of the order in which to combine them.

In fact, the sole input to the decision regarding which tools to use and the order in which to use them is the behavior that is required to be implemented. To this end, our example starts by presenting the network topology and defining the desired behavior. Then we discuss how to combine the tools to match the defined requirements.

The example topology is the simple scenario illustrated in Figure 2.13. A router with one ingress and one egress interface is receiving black and white traffic. The differentiating factor between the two types of packets is the markings they carry. Hence, the markings are the parameter used by the classifier. The PHB requirements to be implemented are summarized in Table 2.1.

Table 2.1 PHB requirements

Traffic	Class of service	Metering and policing			Queue	Scheduler	Egress rate	Rewrite
		Rate	Color	Action				
		<1 M	Green	Accept				
Black	COS1	>5 M < 7 M	Yellow	Accept	Use Q1		Limit	None
		>7 M	Red	Drop		Prioritize	8 Mbps	
		<1 M	Green	Accept		Q1	store excess	
White	COS2	>1 M < 2 M	Yellow	Accept	Use Q2			Marking = X
		>2 M	Red	Drop				None

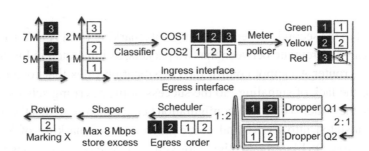

Figure 2.14 Tools combination scenario

Starting with the classification requirements, black packets should be placed in the class of service COS1 and white packets in COS2. Both types of traffic are metered and policed to implement the coloring requirements listed in Table 2.1. In terms of the queuing and scheduling requirements, black packets should be placed in Q1, which has a higher priority on egress, and white packets are placed in Q2. The rate of traffic leaving the router should not exceed 8 megabits per second (Mbps), but the rate control should also ensure that excess traffic is stored whenever possible rather than being discarded.

The final requirement regards the rewrite rules. White packets colored yellow should have their marking set to the value X before leaving the router, to signal this coloring scheme to the next downstream router.

These requirements are all the necessary details regarding the PHB. Let us now combine the QOS tools previously presented to achieve this PHB, as illustrated in Figure 2.14. In Figure 2.14, only three white and three black packets are considered for ease of understanding.

Figure 2.15   Adding a downstream neighbor

The first step is classification, to inspect the packets' markings and identify the traffic. Black traffic is mapped to the class of service COS1 and white traffic to the class of service COS2. From this point onward the traffic is identified, meaning that the behavior of all the remaining tools toward a specific packet is a function of the class of service assigned by the classifier.

The metering and policing tools are the next to be applied. The metering tool colors each packet according to the input arrival rate. The result is that for both the white and black traffic types, one packet is marked green, one marked yellow, and the third marked red. As per the defined requirements, the policer drops the red packet and transmits all the others. The result is that black packet 3 and white packet 3 are discarded.

Now moving to the egress interface, packets are placed into the egress queues, with black packets going to Q1 and white packets to Q2. We assume that because of the current queue fill levels, the dropper associated with each queue allows the packets to be put into the queues. The requirements state that the scheduler should prioritize Q1. As such, the scheduler first transmits packets in Q1 (these are black packets 1 and 2), and only after Q1 is empty does it transmit the packets present in Q2 (white packets 1 and 2). This transmission scheme clearly represents an uneven division of resources applied by the QOS tool to comply with the requirement of prioritizing black traffic.

The shaper block then imposes the desired outbound transmission rate, and when possible, excess traffic is stored inside the shaper.

The last block applied is the rewrite tool, which implements the last remaining requirement that white packets colored yellow should have their marking rewritten to the value X. The rewrite block is usually applied last (but this is not mandatory), following the logic that spending cycles to rewrite the packets markings should occur only once they have passed through other QOS tools that can drop them, such as the policer.

The behavior described previously takes place inside one router. Let us make the example more interesting by adding a downstream neighbor (called router N) that maintains the same set of requirements listed in Table 2.1. For ease of understanding, we again assume that only white and black traffic is present, as illustrated in Figure 2.15.

What set of tools should router N implement to achieve consistency? The same?

The PHB concept implies that once packets leave a router, all the previous classification is effectively lost, so the first mandatory step for router N to take is to classify traffic. It should not only differentiate between white and black packets but should also take into account that packets with a marking of X represent white packets that were colored yellow by its neighbor. Otherwise, having the upstream router use the rewrite tool becomes pointless.

Regarding metering and policing, if the rates defined in Table 2.1 have been imposed by the upstream router, router N does not need to apply such tools. However, this decision touches the topic of trust relationships. If router N cannot trust its neighbor to impose the desired rates, it must apply tools to enforce them itself. Trust relationships are covered in Chapter 3.

Regarding queuing and scheduling, it is mandatory for router N to implement the scheme described previously, because this is the only possible way to ensure that the requirement of prioritizing black traffic is applied consistently across all the routers that the traffic crosses.

Let us now remove router N from the equation and replace it by adding bidirectional traffic, as illustrated in Figure 2.16. Achieving the set of requirements shown in Table 2.1 for the bidirectional traffic in Figure 2.16 is just a question of replicating tools. For example, all the tools we presented previously are applied to interface A, but some, such as metering and policing, are applied only to traffic entering the interface (ingress from the router's perspective), and others, such as shaping, queuing, and scheduling, are applied only to traffic leaving the interface (egress from the router's perspective).

This is a key point to remember; QOS tools are always applied with a sense of directionality, so if, for example, it is required to police both the traffic that enters and leaves an interface, effectively what is applied is one policer in the ingress direction and another policer in the egress direction.

Any QOS tool can be applied at the ingress or egress; the impetus for such decisions is always the desired behavior. This example uses policing as an ingress tool and shaping as an egress tool. However, this design should never be seen as mandatory, and as we will see in Chapter 6, there are scenarios in which these two tools should be applied in the opposite order. The only exception is the

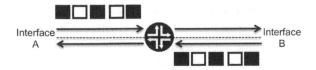

Interface
A

Interface
B

Figure 2.16   Bidirectional traffic

classifier tool. Because the PHB concept implies that any classification done by a router is effectively lost once traffic leaves the router toward its next downstream neighbor, applying it on egress is pointless.

This simple example is drawn from a basic set of requirements. Parts Two and Three of this book show examples of using these tools to achieve greater levels of complexity and granularity.

2.10   Delay and Jitter Insertion

Two QOS tools can insert delay, the shaper and a combination of queuing and scheduling. Because the value of inserted delay is not constant, both tools can also introduce jitter.

The shaper can sustain excess traffic at the expense of delaying it. This means that there is a trade-off between how much excess traffic can be stored and the maximum delay that can possibly be inserted. We explore this topic further in Chapter 6, in which we present the leaky bucket concept used to implement shaping.

Let us now focus on the queuing and scheduling tool. The delay inserted results from a multiplexing operation applied by this QOS tool, as follows. When multiple queues containing packets arrive in parallel at the scheduler, the scheduler selects one queue at a time and removes one or more packets from that queue. While the scheduler is removing packets from the selected queue, all the other packets in this particular queue, as well as all packets in other queues, must wait until it is their turn to be removed from the queue.

Let us illustrate this behavior with the example in Figure 2.17, which shows two queues named A and B and a scheduler that services them in a round-robin fashion, starting by servicing queue A.

Packet X is the last packet inside queue B, and we are interested in calculating the delay introduced into its transmission. The clock starts ticking.

The first action taken by the scheduler is to remove black packet 1 from queue A. Then, because the scheduler is working in a round-robin fashion, it

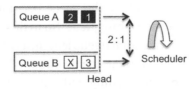

Figure 2.17   Two queues and a round-robin scheduler

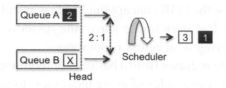

Figure 2.18   Packet X standing at the queue B head

next turns to queue B and removes white packet 3 from the queue. This leads to the scenario illustrated in Figure 2.18, in which packet X is sitting at the head of queue B.

Continuing its round-robin operation, the scheduler now services queue A by removing black packet 2. It again services queue B, which finally results in the removal of packet X. The clock stops ticking.

In this example, the delay introduced into the transmission of packet X is the time that elapses while the scheduler removes black packet 1 from queue A, white packet 3 from queue B, and black packet 2 from queue A.

Considering a generic queuing and scheduling scenario, when a packet enters a queue, the time it takes until the packet reaches the head of the queue depends on two factors, the queue fill level, that is, how many packets are in front of the packet, and how long it takes for the packet to move forward to the head of the queue. For a packet to move forward to the queue head, all the other packets queued in front of it need to be removed first.

As a side note, a packet does not wait indefinitely at the queue head to be removed by the scheduler. Most queuing algorithms apply the concept of packet aging. If the packet is at the queue head for too long, it is dropped. This topic is discussed in Chapter 7.

The speed at which packets are removed from the queue is determined by the scheduler's properties regarding how fast and how often it services the queue, as shown with the example in Figures 2.17 and 2.18. But for now, let us concentrate only on the queue fill level, and we will return to the scheduler's removal speed shortly.

It is not possible to predict the queue fill level, so we have to focus on the worst-case scenario of a full queue, which allows us to calculate the maximum delay value that can be inserted into the packet's transmission. When a packet enters a queue and when this results in the queue becoming full, this particular packet becomes the last one inside a full queue. So in this case, the delay inserted in that packet's transmission is the total queue length, as illustrated in Figure 2.19.

To summarize our conclusions so far, the maximum delay that can be inserted in a packet's transmission by the queuing and scheduling tools is the length of the queue into which the packet is placed.

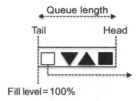

Figure 2.19   Worst-case delay scenario with a full queue

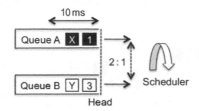

Figure 2.20   Two equal queues

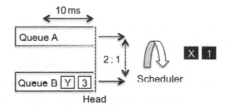

Figure 2.21   Uneven scheduler operation

Now if we take into account the role played by the scheduler in determining the speed at which packets are removed, the conclusion drawn in the previous paragraph becomes a reasonable approximation of the delay.

Let us demonstrate how the accuracy of this conclusion varies according to the scheduler's properties. The example in Figure 2.20 shows two queues that are equal in the sense that both have a length of 10 ms, each contain two packets, and both are full.

Packets X and Y are the last packets in the full queues A and B, respectively. In this case, the scheduler property that is implemented is that as long as packets are present in queue A, this queue is always serviced first. The result of this behavior leads the scheduler to first remove black packets 1 and X. Only then does it turn to queue B, leading to the scenario illustrated in Figure 2.21.

This example shows that the previous conclusion drawn—that the maximum delay that can be inserted by the queuing and scheduling tools is the length of the queue on which the packet is placed—is much more accurate for packet X than for packet Y, because the scheduler favors the queue A at the expense of penalizing queue B.

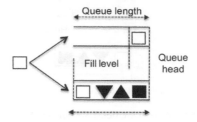

Figure 2.22   Best-case and worst-case scenarios in terms of jitter insertion

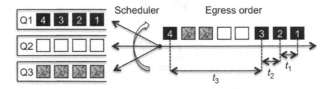

Figure 2.23   Jitter insertion due to scheduler jumps

So, for packets using queue A, to state that the maximum possible delay that can be inserted is 10 ms (the queue length) is a good approximation in terms of accuracy. However, this same statement is less accurate for queue B.

In a nutshell, the queue size can be seen as the maximum amount of delay inserted, and this approximation becomes more accurate when the scheduling scheme favors one particular queue and becomes less accurate for other queues.

Regarding jitter, the main factors that affect it are also the queue length and the queue fill level, along with a third phenomenon called the scheduler jumps that we discuss later in this section. As the gap between introducing no delay and the maximum possible value of delay widens, the possible value of jitter inserted also increases.

As illustrated in Figure 2.22, the best-case scenario for the delay parameter is for the packet to be mapped into an empty queue. In this case, the packet is automatically placed at the queue head, and it has to wait only for the scheduler to service this queue. The worst-case scenario has already been discussed, one in which the packet entering a queue is the last packet that the queue can accommodate. The result is that as the queue length increases, the maximum possible variation of delay (jitter) increases as well.

Scheduler jumps are a consequence of having multiple queues, so the scheduler services one queue and then needs to jump to service other queues. Let us illustrate the effect such a phenomenon has in terms of jitter by considering the example in Figure 2.23, in which a scheduler services three queues.

In its operation, the scheduler removes three packets from Q1 and then two packets from Q2 and other two packets from Q3. Only then does it jump again to Q1 to remove packets from this queue. As illustrated in Figure 2.23, the implications in terms of jitter is that the time elapsed between the transmission of black packets 2 and 3 is smaller than the time elapsed between the transmission of black packets 3 and 4. This is jitter. Scheduler jumps are inevitable, and the only way to minimize them is to use the minimum number of queues, but doing so without compromising the traffic differentiation that is achieved by splitting traffic into different queues.

As discussed in Chapter 8, a queue that carries real-time traffic typically has a short length, and the scheduler prioritizes it with regards to the other queues, meaning it removes more packets from this queue compared with the others to minimize the possible introduction of delay and jitter. However, this scheme should be achieved in a way that assures that the other queues do not suffer from complete resource starvation.

2.11   Packet Loss

At a first glance, packet loss seems like something to be avoided, but as we will see, this is not always the case because in certain scenarios, it is preferable to drop packets rather than transmit them.

Three QOS tools can cause packet loss: the policer, the shaper, and the queuing. From a practical perspective, QOS packet loss tools can be divided into two groups depending on whether traffic is dropped because of an explicitly defined action or is dropped implicitly because not enough resources are available to cope with it.

The policer belongs to the first group. When traffic exceeds a certain rate and if the action defined is to drop it, traffic is effectively dropped and packet loss occurs. Usually, this dropping of packets by the policer happens for one of two reasons: either the allowed rate has been inaccurately dimensioned or the amount of traffic is indeed above the agreed or expected rate and, as such, it should be dropped.

The shaper and queuing tools belong to the second group. They drop traffic only when they run out of resources, where the term *resources* refers to the maximum amount of excess traffic that can be sustained for the shaper and the queue length for the queuing tool (assuming that the dropper drops traffic only when the fill level is 100%).

As previously discussed, there is a direct relationship between the amount of resources assigned to the shaper and queuing tools (effectively, the queue length) and the maximum amount of delay and jitter that can be inserted. Thus,

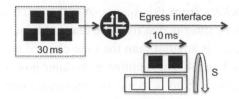

Figure 2.24   Traffic dropped due to the lack of queuing resources

limiting the amount of resources implies lowering the maximum values of delay and jitter that can be inserted, which is crucial for real-time traffic because of its sensitivity to these parameters. However, limiting the resources has the side effect of increasing the probability of dropping traffic.

Suppose a real-time traffic stream crossing a router has the requirement that a delay greater than 10 ms is not acceptable. Also suppose that inside the router, the real-time traffic is placed in a specific egress queue whose length is set to less than 10 ms to comply with the desired requirements. Let us further suppose that in a certain period of time, the amount of traffic that arrives at the router has an abrupt variation in volume, thus requiring 30 ms worth of buffering, as illustrated in Figure 2.24.

The result is that some traffic is dropped. However, the packet loss introduced can be seen as a positive situation, because we are not transmitting packets that violate the established service agreement. Also, as previously discussed, for a real-time stream, a packet that arrives outside the time window in which it is considered relevant not only adds no value but also causes more harm than good because the receiver must still spend cycles processing an already useless packet.

This perspective can be expanded to consider whether it makes sense to even allow this burst of 30 ms worth of traffic to enter the router on an ingress interface if, on egress, the traffic is mapped to a queue whose length is only 10 ms. This is a topic that we revisit in Chapter 6, in which we discuss using the token bucket concept to implement policing and to deal with traffic bursts.

2.12   Conclusion

The focus of this chapter has been to present the QOS tools as building blocks, where each one plays a specific role in achieving various goals. As the demand for QOS increases, the tools become more refined and granular. As an example, queuing and scheduling can be applied in a multiple level hierarchical manner. However, the key starting point is understanding what queuing and scheduling can achieve as building blocks in the QOS design.

Some tools are more complex than others. Thus, in Part Two of this book we dive more deeply into the internal mechanics of classifiers, policers, and shapers and the queuing and scheduling schemes.

We have also presented an example of how all these tools can be combined. However, the reader should always keep in mind that the required tools are a function of the desired end goal.

In the next chapter, we focus on some of the challenges and particulars involved in a QOS deployment.

Reference

[1]   Heinanen, J., Finland, T. and Guerin, R. (1999) RFC 2698, A Two-Rate Three-Color Marker, September 1999. https://tools.ietf.org/html/rfc2698 (accessed August 19, 2015).

3

Challenges

In the previous chapter, we discussed the QOS toolkit that is available as part of a QOS deployment on a router. We now move on, leaving behind the perspective of an isolated router and considering a network-wide QOS deployment. Such deployments always have peculiarities depending on the business requirements, which make each single one unique. However, the challenges that are likely to be present across most deployments are the subject of this chapter.

Within a QOS network and on each particular router in the network, multiple traffic types compete for the same network resources. The role of QOS is to provide each traffic type with the behavior that fits its needs. So the first challenge that needs to be considered is how providing the required behavior to a particular traffic type will have an impact and will place limits on the behavior that can be offered to the other traffic types. As discussed in Chapter 2, the inherent delay in the queuing and scheduling operation can be minimized for traffic that is placed inside a particular queue. However, that is achieved at the expense of increasing the delay for the traffic present in other queues. The unavoidable fact that something will be penalized is true for any QOS tool that combines a greater number of inputs into a fewer number of outputs.

This description of QOS behavior can also be stated in a much more provocative way: a network in which all traffic is equally "very important and top priority" has no room for QOS.

QOS-Enabled Networks: Tools and Foundations, Second Edition. Miguel Barreiros and Peter Lundqvist.
© 2016 John Wiley & Sons, Ltd. Published 2016 by John Wiley & Sons, Ltd.

3.1   Defining the Classes of Service

The main foundation of the entire QOS concept is applying different behavior to different traffic types. Achieving traffic differentiation is mandatory, because it is only by splitting traffic into different classes of service that different behavior can be selectively applied to each.

In Chapter 2, we presented the classifier tool that, based on its own set of rules (which are explored further in Chapter 5), makes decisions regarding the class of service to which traffic belongs. Let us now discuss the definition of the classes of service themselves.

In the DiffServ model, each router first classifies traffic and then, according to the result of that classification, applies a specific per-hop behavior (PHB) to it. Consistency is achieved by ensuring that each router present along the path that the traffic takes across the network applies the same PHB to the traffic.

A class of service represents a traffic aggregation group, in which all traffic belonging to a specific class of service has the same behavioral requirements in terms of the PHB that should be applied. This concept is commonly called a behavior aggregate.

Let us return to the example we presented in Chapter 2, when we illustrated the combination of QOS tools by using two classes of service named COS1 and COS2. Traffic belonging to COS1 was prioritized in terms of queuing and scheduling to lower the delay inserted in its transmission, as illustrated again in Figure 3.1.

In this case, the router can apply two different behaviors in terms of the delay that is inserted in the traffic transmission, and the classifier makes the decision regarding which one is applied when it maps traffic into one of the two available classes of service. So continuing this example, on one side of the equation we have traffic belonging to different services or applications with their requirements, and on the other, we have the two classes of service, each of which corresponds to a specific PHB, which in this case is characterized by the amount of delay introduced, as illustrated in Figure 3.2.

The relationship between services or applications and classes of service should be seen as N : 1, not as 1 : 1, meaning that traffic belonging to different

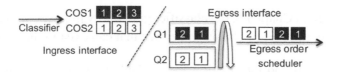

Figure 3.1   Prioritizing one class of service

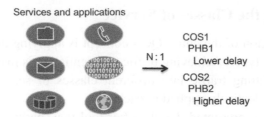

Figure 3.2   Mapping between services and classes of service

services or applications but with the same behavior requirements should be mapped to the same class of service. For example, two packets belonging to two different real-time services but having the same requirements in terms of the behavior they should receive from the network should be classified into the same class of service. The only exception to this is network control traffic, as we see later in this chapter.

The crucial question then becomes how many and what different behaviors need to be implemented. As with many things in the QOS realm, there is no generic answer, because the business drivers tend to make each scenario unique.

Returning to Figure 3.2, the approach is to first identify the various services and applications the network needs to support, and then take into account any behavior requirements, and similarities among them, to determine the number of different behaviors that need to be implemented.

Something commonly seen in the field is the creation of as many classes of service as possible. Conceptually, this is the wrong approach. The approach should indeed be the opposite to create only the minimum number of classes of service. There are several reasons behind this logic:

- The more different behaviors the network needs to implement, the more complex it becomes, which has implications in terms of network operation and management.
- As previously stated, QOS does not make the road wider, so although traffic can be split into a vast number of classes of service, the amount of resources available for traffic as a whole remains the same.
- The number of queues and their length are limited (a topic discussed later in this chapter).
- As we will see in Chapter 5, the classifier granularity imposes limits regarding the maximum number of classes of service that can exist in the network.

Plenty of standards and information are available in the networking world that can advise the reader on what classes of service should be used, and some even name suggestions. While this information can be useful as a guideline, the reader should view them critically because a generic solution is very rarely appropriate for a particular scenario. That is the reason why this book offers no generic recommendations in terms of the classes of service that should exist in a network.

Business drivers shape the QOS deployment, and not the other way round, so only when the business drivers are present, as in the case studies in Part Three of this book, do the authors provide recommendations and guidelines regarding the classes of service that should be used.

3.2   Classes of Service and Queues Mapping

As presented in Chapter 2, the combination of the queuing and scheduling tools directs traffic from several queues into a single output, and the queue properties, allied with the scheduling rules, dictate specific behavior regarding delay, jitter, and packet loss, as illustrated in Figure 3.3.

As also discussed in Chapter 2, other tools can have an impact in terms of delay, jitter, and packet loss. However, the queuing and scheduling stage is special in the sense that it is where the traffic from different queues is combined into a single output.

So if, after taking into account the required behavior, traffic is aggregated into classes of service, and if each queue associated with the scheduling policy provides a specific behavior, then mapping each class of service to a specific queue is recommended. A 1:1 mapping between queues and classes of service aligns with the concept that traffic mapped to each class of service should receive a specific PHB.

Also, if each class of service is mapped to a unique queue, the inputs for the definition of the scheduler rules that define how it serves the queues should themselves be the class-of-service requirements.

When we previously discussed creation of the classes of service, we considered that all traffic classified into a specific class of service has the same behavior requirements. However, as we saw in Chapter 2, the application of the CIR/PIR model can differentiate among traffic inside one class of service. A 1:1

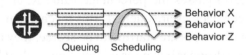

Queuing   Scheduling

Figure 3.3   Each queue associated with the scheduling policy provides a specific behavior

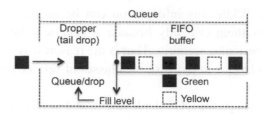

Figure 3.4   Green and yellow traffic in the same queue

mapping between queues and classes of service can become challenging if some traffic in a queue is green (in contract) and other traffic is yellow (out of contract). The concern is how to protect resources for green traffic. Figure 3.4 illustrates this problem, showing a case in which both green and yellow traffic are mapped to the same queue and this queue is full.

As shown in Figure 3.4, the queue is full with both green and yellow packets. When the next packet arrives at this queue, the queue is indifferent as to whether the packet is green or yellow, and the packet is dropped. Because yellow packets inside the queue are consuming queuing resources, any newly arrived green packets are discarded because the queue is full. This behavior is conceptually wrong, because as previously discussed, the network must protect green traffic before accepting yellow traffic.

There are two possible solutions for this problem. The first is to differentiate between green and yellow packets within the same queue. The second is to use different queues for green and yellow packets and then differentiate at the scheduler level.

Let us start by demonstrating how to differentiate between different types of traffic within the same queue. The behavior shown in Figure 3.4 is called *tail drop*. When the queue fill level is at 100%, the dropper block associated with the queue drops all newly arrived packets, regardless of whether they are green or yellow. To achieve differentiation between packets according to their color, the dropper needs to be more granular so that it can apply different drop probabilities based on the traffic color. As exemplified in Figure 3.5, the dropper block can implement a behavior such that once the queue fill level is at X% (or goes above that value), no more yellow packets are accepted in the queue, while green packets are dropped only when the queue is full (fill level of 100%).

Comparing Figures 3.4 and 3.5, the striking difference is that, in Figure 3.5, once the queue fill level passes the percentage value X, all yellow packets are dropped and only green packets are queued. This mechanism defines a threshold so when the queuing resources are starting to be scarce, they are accessible only for green packets. This dropper behavior is commonly called Weighted Random Early Discard (WRED), and we provide details of this in Chapter 8.

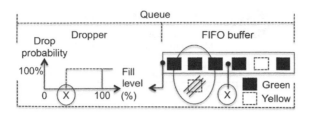

Figure 3.5   Different dropper behaviors applied to green and yellow traffic

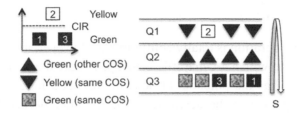

Figure 3.6   Using a different queue for yellow traffic

The second possible solution is to place green and yellow traffic in separate queues and then differentiate using scheduling policy. This approach conforms with the concept of applying a different behavior to green and yellow traffic. However, it comes with its own set of challenges.

Let us consider the scenario illustrated in Figure 3.6, in which three sequential packets, numbered 1 through 3 and belonging to the same application, are queued. However, the metering and policing functionality marks the second packet, the white one, as yellow.

As per the scenario of Figure 3.6, there are three queues in which different types of traffic are mapped. Q1 is used for out-of-contract traffic belonging to this and other applications, so packet 2 is mixed with yellow packets belonging to the same class of service, represented in Figure 3.6 as inverted triangles. Q2 is used by green packets of another class of service. Finally, Q3 is used by packets 1 and 3 and also by green packets that belong to the same class of service.

So we have green and yellow packets placed in different queues, which ensures that the scenario illustrated in Figure 3.4, in which a queue full with green and yellow packets leads to tail dropping of any newly arrived green packet, is not possible. However, in solving one problem we are potentially creating another.

The fact that green and yellow traffic is placed into two different queues can lead to a scenario in which packets arrive at the destination out of sequence. For example, packet 3 can be transmitted before packet 2.

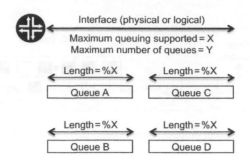

Figure 3.7   Maximum number of queues and maximum length

The scheduler operation is totally configurable. However, it is logical for it to favor Q2 and Q3, to which green traffic is mapped, more than Q1 which, returning to Figure 3.6, has the potential of delaying packet 2 long enough for it to arrive out of sequence at the destination, that is, after packet 3 has arrived. Packet reordering issues can be prevented only if traffic is mapped to the same queue, because, as explained in Chapter 2, packets cannot overtake each other within a queue.

The choice between the two solutions presented above is a question of analyzing the different drawbacks of each. Using WRED increases the probability of dropping yellow traffic, and using a different queue increases the probability of introducing packet reordering issues at the destination.

Adding to the above situation, the queuing resources—how many queues there are and their maximum length—are always finite numbers, so dedicating one queue to carry yellow traffic may pose a scaling problem as well. An interface can support a maximum number of queues, and the total sum of the queue lengths supported by an interface is also limited to a maximum value (we will call it X), as exemplified in Figure 3.7 for a scenario of four queues.

The strategy of "the more queues, the better" can have its drawbacks because, besides the existence of a maximum number of queues, the value X must be divided across all the queues that exist on the interface. Suppose a scenario of queues A and B, each one requiring 40% of the value X. By simple mathematics, the sum of the lengths of all other remaining queues is limited to 20% of X, which can be a problem if any other queue also requires a large length.

3.3   Inherent Delay Factors

When traffic crosses a network from source to destination, the total amount of delay inserted at each hop can be categorized in smaller factors that contribute to the overall delay value.

There are two groups of contributors to delay. The first group encompasses the QOS tools that insert delay due to their inherent operation, and the second group inserts delay as a result of the transmission of packets between routers. While the second group is not directly linked to the QOS realm, it still needs to be accounted for.

For the first group, and as discussed in Chapter 2, two QOS tools can insert delay: the shaper and the combination of queuing and scheduling.

Let us now move to the second group. The transmission of a packet between two adjacent routers is always subject to two types of delay: serialization and propagation.

The serialization delay is the time it takes at the egress interface to place a packet on the link toward the next router. An interface bandwidth is measured in bits per second, so if the packet length is X bits, the question becomes, how much time does it take to serialize it? The amount of time depends on the packet length and the interface bandwidth.

The propagation delay is the time the signal takes to propagate itself in the medium that connects two routers. For example, if two routers are connected with an optical fiber, the propagation delay is the time it takes the signal to travel from one end to the other inside the optical fiber. It is a constant value for each specific medium.

Let us give an example of how these values are combined with each other by using the scenario illustrated in Figure 3.8. In Figure 3.8, at the egress interface, the white packet is placed in a queue, where it waits to be removed from the queue by the scheduler. This wait is the first delay value to account for. Once the packet is removed by the scheduler, and assuming that no shaping is applied, the packet is serialized onto the wire, which is the second delay value that needs to be accounted for.

Once the packet is on the wire, we must consider the propagation time, the time it takes the packet to travel from the egress interface on this router to the

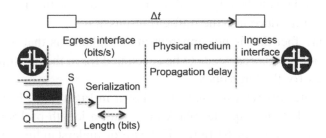

Figure 3.8   Delay incurred at each hop

ingress interface on next router via the physical medium that connects both routers. This time is the third and last delay value to be accounted for.

We are ignoring a possible fourth value, namely, the processing delay, because we are assuming the forwarding and routing planes of the router are independent and that the forwarding plane does not introduce any delay into the packet processing.

The key difference between these two groups of contributors to delay factors is control. Focusing on the first group, the delay inherent in the shaping tool is applied only to packets that cross the shaper, where it is possible to select which classes of service are shaped. The presence of queuing and scheduling implies the introduction of delay, but how much delay is inserted into traffic belonging to each class of service can be controlled by dimensioning the queue lengths and by the scheduler policy. Hence, the QOS deployment offers control over when and how such factors come into play. However, in the second group of delay factors, such control does not exist. Independently of the class of service that the packets belong to, the serialization and propagation delays always exist, because such delay factors are inherent in the transmission of the packets from one router to another.

Serialization delay is dependent on the interface bandwidth and the specific packet length. For example, on an interface with a bandwidth of 64 kbps, the time it takes to serialize a 1500-byte packet is around 188 ms, while for a gigabit interface, the time to serialize the same packet is approximately 0.012 ms.

So for two consecutive packets with lengths of 64 and 1500 bytes, the serialization delay is different for each. However, whether this variation can be ignored depends on the interface bandwidth. For a large bandwidth interfaces, such as a gigabit interface, the difference is of the order of microseconds. However, for a low-speed interface such as one operating at 64 kbps, the difference is of the level of a few orders of magnitude.

Where to draw the boundary between a slow and a fast interface can be done in terms of defining when the serialization delay starts to be able to be ignored, a decision that can be made only by taking into account the maximum delay that is acceptable to introduce into the traffic transmission.

However, one fact that is not apparent at first glance is that a serialization delay that cannot be ignored needs to be accounted for in the transmission of all packets, not just for the large ones.

Let us illustrate this with the scenario shown in Figure 3.9. Here, there are two types of packets in a queue: black ones with a length of 1500 bytes and white ones with a length of 64 bytes. For ease of understanding, this scenario assumes that only one queue is being used.

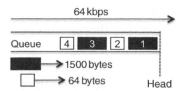

Figure 3.9   One queue with 1500-byte and 64-byte packets

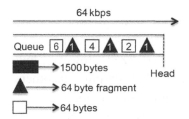

Figure 3.10   Link fragmentation and interleaving operation

Looking at Figure 3.9, when packet 1 is removed from the queue, it is transmitted by the interface in an operation that takes around 188 ms, as explained above. The next packet to be removed is number 2, which has a serialization time of 8 ms. However, this packet is transmitted only by the interface once it has finished serializing packet 1, which takes 188 ms, so a high serialization delay impacts not only large packets but also small ones that are transmitted after the large ones.

There are two possible approaches to solve this problem. The first is to place large packets in separate queues and apply an aggressive scheduling scheme in which queues with large packets are served only when the other ones are empty. This approach has its drawbacks because it is possible that resource starvation will occur on the queues with large packets. Also, this approach can be effective only if all large packets can indeed be grouped into the same queue (or queues).

The second approach consists of breaking the large packets into smaller ones using a technique commonly named link fragmentation and interleaving (LFI).

The only way to reduce a packet serialization delay is to make the packets smaller. LFI fragments the packet into smaller pieces and transmits those fragments instead of transmitting the whole packet. The router at the other end of the link is then responsible for reassembling the packet fragments.

The total serialization time for transmitting the entire packet or for transmitting all its fragments sequentially is effectively the same, so fragmenting is only half of the solution. The other half is interleaving: the fragments are transmitted interleaved between the other small packets, as illustrated in Figure 3.10.

In Figure 3.10, black packet 1 is fragmented into 64-byte chunks, and each fragment is interleaved between the other packets in the queue. The new packet that stands in front of packet 2 has a length of 64 bytes, so it takes 8 ms to serialize instead of the 188 ms required for the single 1500-byte packet.

The drawback is that the delay in transmitting the whole black packet increases, because between each of its fragments other packets are transmitted. Also, the interleaving technique is dependent on support from the next downstream router.

Interface bandwidth values have greatly increased in the last few years, reaching a point where even 10 gigabit is starting to be common, so a 1500-byte packet takes just 1 μs to be serialized. However, low-speed interfaces are still found in legacy networks and in certain customers' network access points, coming mainly from the Frame Relay realm.

The propagation delay is always a constant value that depends on the physical medium. It is typically negligible for connections made using optical fiber or unshielded twisted pair (UTP) cables and usually comes into play only for connections established over satellite links.

The previous paragraphs described the inherent delay factors that can exist when transmitting packets between two routers. Let us now take broader view, looking at a source-to-destination traffic flow across a network. Obviously, the total delay for each packet depends on the amount of delay inserted at each hop. If several possible paths exist from source to destination, it is possible to choose which delay values will be accounted for.

Let us illustrate this using the network topology in Figure 3.11 where there are two possible paths between router one (R1) and two (R2), one is direct and there is a second one that crosses router three (R3). The delay value indicated at each interconnection represents the sum of propagation, serialization, and queuing and scheduling delays at those points.

As shown in Figure 3.11, if the path chosen is the first one, the total value of delay introduced is Delay1. But if the second is chosen, the total value of delay introduced is the sum of the values Delay2 and Delay3.

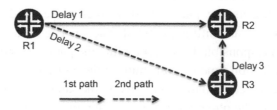

Figure 3.11   Multiple possible paths with different delay values

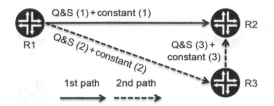

Figure 3.12 Queuing and scheduling delay as the only variable

At a first glance, crossing more hops from source to destination can be seen as negative in terms of the value of delay inserted. However, this is not always the case because it can, for example, allow traffic to avoid a lower-bandwidth interface or a slow physical medium such as a satellite link. What ultimately matters is the total value of delay introduced, not the number of hops crossed, although a connection between the two values is likely to exist.

The propagation delay is constant for each specific medium. If we consider the packet length to be equal to either an average expected value or, if that is unknown, to the link maximum transmission unit (MTU), the serialization delay becomes constant for each interface speed, which allows us to simplify the network topology shown in Figure 3.11 into the one in Figure 3.12, in which the only variable is the queuing and scheduling delay (Q&S).

In an IP or an MPLS network without traffic engineering, although routing is end to end, the forwarding decision regarding how to reach the next downstream router is made independently at each hop. The operator has the flexibility to change the routing metrics associated with the interfaces to reflect the more or less expected delay that traffic faces when crossing them. However, this decision is applied to all traffic without granularity, because all traffic follows the best routing path, except when there are equal-cost paths from the source to the destination.

In an MPLS network with traffic engineering (MPLS-TE), traffic can follow several predefined paths from the source to the destination, including different ones from the path selected by the routing protocol. This flexibility allows more granularity to decide which traffic crosses which hops or, put another way, which traffic is subject to which delay factors. Let us consider the example in Figure 3.13 of two established MPLS-TE LSPs, where LSP number 1 follows the best path as selected by the routing protocol and LSP number 2 takes a different path across the network.

The relevance of this MPLS-TE characteristic in terms of QOS is that it allows traffic to be split into different LSPs that cross different hops from the source to the destination. So, as illustrated in Figure 3.13, black packets are mapped into

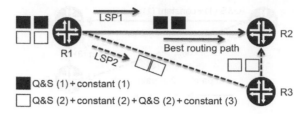

Figure 3.13   Using MPLS-TE to control which traffic is subject to which delay

LSP1 and are subject to the delay value of Q&S (1) plus constant (1), and white packets mapped into LSP2 are subject to a different end-to-end delay value.

3.4   Congestion Points

As previously discussed in Chapter 1, the problem created by the network convergence phenomena is that because different types of traffic with different requirements coexist in the same network infrastructure, allowing them to compete freely does not work. The first solution was to make the road wider, that is, to have so many resources available that there would never be any resource shortage. Exaggerating the road metaphor and considering a street with 10 houses and 10 different lanes, even if everybody leaves for work at the same time, there should never be a traffic jam. However, this approach was abandoned because it goes against the main business driver for networks convergence, that is, cost reduction.

Congestion points in the network exist when there is a resource shortage, and the importance of QOS within a network increases as the available network resources shrink.

An important point to be made yet again is that QOS does not make the road wider. For example, a gigabit interface with or without QOS has always the same bandwidth, 1 gigabit. A congestion point is created when the total amount of traffic targeted for a destination exceeds the available bandwidth to that destination, for example, when the total amount of traffic exceeds the physical interface bandwidth, as illustrated in Figure 3.14.

Also, a congestion scenario can be artificially created, for example, when the bandwidth contracted by the customer is lower than the physical interface bandwidth, as illustrated in Figure 3.15.

At a network congestion point, two QOS features are useful: delay and prioritization. Delay can be viewed as an alternative to dropping traffic, holding the traffic back until there are resources to transmit it. As we saw in Chapter 2, both

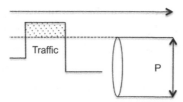

Figure 3.14 Congestion point because the traffic rate is higher than the physical interface bandwidth. P, physical interface bandwidth

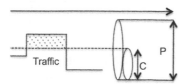

Figure 3.15 Congestion point artificially created. C, contracted bandwidth; P, physical interface bandwidth

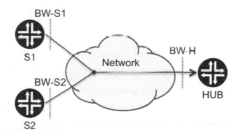

Figure 3.16 Congestion point in a hub-and-spoke topology

the shaper and queuing tools are able to store traffic. Delay combined with prioritization is the role played by the combination of queuing and scheduling. That is, the aim is to store the traffic, to be able to select which type of traffic is more important, and to transmit that type first or more often. The side effect is that other traffic types have to wait to be transmitted until it is their turn.

But let us use a practical scenario for a congestion points, the hub-and-spoke topology. Figure 3.16 shows two spoke sites, named S1 and S2, which communicate with each other via the hub site. The interfaces' bandwidth values between the network and the sites S1, S2, and the hub are called BW-S1, BW-S2, and BW-H, respectively.

Dimensioning of the BW-H value can be done in two different ways. The first approach is the "maximum resources" one: just make BW-H equal to the sum of BW-S1 and BW-S2. With this approach, even if the two spoke sites are transmitting at full rate to the hub, there is no shortage of bandwidth resources.

The second approach is to use a smaller value for BW-H, following the logic that situations when both spoke sites are transmitting at full rate to the hub will be transient. However, when those transient situations do happen, congestion will occur, so QOS tools will need to be set in motion to avoid packets being dropped. The business driver here is once again cost: requiring a lower bandwidth value is bound to have an impact in terms of cost reduction.

In the previous paragraph, we used the term "transient," and this is an important point to bear in mind. The amount of traffic that can be stored inside any QOS tool is always limited, so a permanent congestion scenario unavoidably leads to the exhaustion of the QOS tool's ability to store traffic, and packets will be dropped.

The previous example focuses on a congestion point on a customer-facing interface. Let us now turn to inside of the network itself. The existence of a congestion point inside the network is usually due to a failure scenario, because when a network core is in a steady state, it should not have any bandwidth shortage. This is not to say, however, that QOS tools are not applicable, because different types of traffic still require different treatment. For example, queuing and scheduling are likely to be present.

Let us use the example illustrated in Figure 3.17, in which, in the steady state, all links have a load of 75% of their maximum capability. However, a link failure between routers R1 and R2 creates a congestion point in the link named X, and that link has queuing and scheduling enabled. A congestion point is created on the egress interface on R1.

Traffic prioritization still works, and queues in which traffic is considered to be more important are still favored by the scheduler. However, the queuing part of the equation is not that simple. The amount of traffic that can be stored inside a particular queue is a function of its length, so the pertinent question is, should queues be dimensioned for the steady-state scenario or should they also take

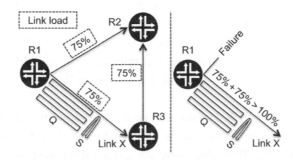

Figure 3.17   Congestion point due to a failure scenario

into account possible failure scenarios? Because we still have not yet presented all the pieces of the puzzle regarding queuing and scheduling, we will analyze this question in Chapter 8.

As a result of the situation illustrated in Figure 3.17, most operators these days use the rule, "if link usage reaches 50%, then upgrade." This rule follows the logic that if a link fails, the "other" link has enough bandwidth to carry all the traffic, which is a solution to the problem illustrated in Figure 3.17. However, considering a network not with three but with hundreds of routers, implementing such logic becomes challenging. Here the discussion moves somewhat away from the QOS realm and enters the MPLS world of being able to have predictable primary and secondary paths between source and destination. This is another topic that we leave for Chapter 8 and for the case studies in Part Three.

3.5 Trust Borders

When traffic arrives at a router, in terms of QOS, the router can trust it or not. The term trust can be seen from two different perspectives. The first is whether the information present in the packets is valid input for the classifier deciding the class of service to which the packets belong. The concern is assuring that the classifier is not fooled by any misinformation present in the packets, which could lead to traffic being placed in an incorrect class of service.

The second perspective is whether an agreement has been established between both parties regarding the amount of resources that should be used at the interconnection point. The question here is whether the router can trust the other side to keep its part of the agreement or whether the router needs to enforce it.

If two routers belong to the same network, it is expected that the downstream router can trust the traffic it receives from its neighbor and also that the neighbor is complying with any agreement that has been established. However, this is not the case at the border between networks that belong to different entities.

A trust border is a point where traffic changes hands, moving from using the resources of one network to using those of another network. The term "different" should be seen from a control perspective, not from a topology perspective, because it is perfectly possible for the same entity to own two different networks and hence to be able to trust the information they receive from each other at the interconnection points.

An agreement regarding resources usage usually is in place at trust borders. However, the network that receives the traffic cannot just assume that the other side is keeping its part of the agreement; it needs to be sure of it. So two conditions need to be imposed: first, any traffic that should not be allowed in the network because it violates the agreement should get no farther, and, second, traffic needs to be trusted before being delivered to the next downstream router, effectively before entering the network trust zone.

Let us illustrate the points in the previous paragraphs with a practical example. Assume that the service contracted by a customer from the network is a total bandwidth of 10 Mbps (megabits per second) and that two classes of service have been hired, voice and Internet, where voice packets are identified by having the marking X. Bandwidth in the voice class is more expensive because of assurances of lower delay and jitter, so from the total aggregate rate of 10 Mbps hired, the customer has bought only 2 Mbps of voice traffic.

Let us assume that the customer is violating the agreement by sending 20 Mbps of traffic and all packets are marked X, a situation which, if left unchecked, can lead to the usage of network resources that are not part of the agreement (see Figure 3.18).

Before allowing traffic that is arriving from the customer to enter the network trust zone, the border router needs to make it conform to the agreement made by limiting it to 10 Mbps and by changing the markings of packets above the rate hired for voice traffic.

The change of the packets markings is necessary because, following the PHB concept, the next downstream router applies the classifier tool, and packets with the wrong marking inside the trust zone have the potential of fooling the classifier and jumping to a class of service to which they should not have access. Another possible option is for the border router to simply drop the packets that contain an incorrect marking.

An interesting scenario is the one of a layered network, for example, MPLS VPNs, in which traffic is received from the customer as Ethernet or IP and transits through the network encapsulated inside MPLS packets. Here, the information contained in the Ethernet or IP packets header is not relevant inside the network, because the routers inspect only the MPLS header. We explore such a scenario in the case studies in Part Three of this book.

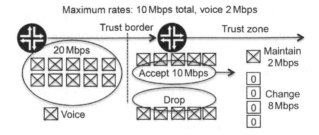

Figure 3.18   The border between trust zones

3.6   Granularity Levels

Routers inside a network fall into two main groups: routers that are placed at the edge of the network, commonly named provider edge (PE), and core routers, commonly named provider (P). From a QOS perspective, the key differentiation factor between PE and P routers is not their position in the network topology but the types of interfaces they have. A PE router has two types of interfaces, customer and core facing, while P routers only have the second type.

A core-facing interface is where the connection between two network routers is made, and a customer-facing interface is where the service end points are located. A core- or customer-facing interface is not required to be a physical interface. It can also be a logical interface such as, for example, a specific VLAN in a physical Ethernet interface.

The division between interface types is interesting because of the granularity levels that need to be considered at each stage. Let us start by considering the example in Figure 3.19, which shows three customers, numbered 1 through 3, connected to the PE router, on two physical customer-facing interfaces (P1 and P2), and where traffic flows to or from the other network router via a core-facing interface.

Customer 1 uses service A and customer 2 uses service B, and both services are classified into COS1. Customer 3 uses service C, which is classified into COS2. The physical interface P1 has two logical interfaces, L1 and L2, on which the service end points for customers 1 and 2, respectively, are located. The interest in differentiating between types of interfaces is being able to define the required granularity levels.

All traffic present in the network should always be classified to ensure that there is never any doubt regarding the behavior that should be applied to it. As such, the lowest level of granularity that can exist is to simply identify traffic as belonging to one class of service, which encompasses all customer services

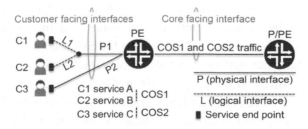

Figure 3.19   Customer- and core-facing interfaces

and any other network traffic mapped to that class of service. A higher level of granularity would be to identify traffic as belonging to a certain class of service and also to a specific customer.

A core-facing interface has no service end points because it operates as a transit point, where traffic belonging to multiple classes of service flows through it. As such, the granularity level required is usually the lowest one. A core-facing interface does not need to know which particular customer the traffic belongs to. Returning to Figure 3.19, the core interface has no need to be concerned if the traffic mapped into COS1 that crosses it belongs to service A or B or to customer 1 or 2. All it should be aware of is the class of service to which the traffic belongs, because, in principle, traffic from two customers mapped to the same class of service should receive the same behavior from the network. However, the core interface always needs to be able to differentiate between the different classes of service, COS1 and COS2, to be able to apply different behaviors to each.

As for customer-facing interfaces, the granularity level required is usually the highest. In Figure 3.19, two service end points are located on physical interface 1, so just knowing the class of service is not enough. It is also necessary to know the specific customer to which the traffic belongs. An interesting scenario is that of customer 3, who possesses the only service end point located on physical interface 2. For this particular scenario of a single service end point on one physical interface, the interface itself identifies the customer so the granularity level required can be the lowest.

It is important to identify the necessary granularity levels at each interface because this has a direct bearing on how granular the QOS tools need to be. In Figure 3.19, in the core interface, traffic belonging to the class of service COS1 can all be queued together. However, on the customer-facing interface 1, queuing should be done on a per-customer basis, because the same interface has multiple service end points.

In a nutshell, core-facing interfaces typically should have the lowest granularity level (i.e., these interfaces should be aware only of class of service) and customer-facing interfaces should have the highest granularity level (i.e., these interfaces should be aware of both the customer and the class of service). However, exceptions can exist, as highlighted for customer 3.

When designing the network and deciding which tools to apply on each interface, the required granularity levels make a difference when selecting which QOS tools are necessary and how granular they need to be. These choices are typically closely tied with the hardware requirements for each type of interface.

3.7   Control Traffic

So far, we have focused the discussion on QOS on customer traffic that transits the network between service end points. But there is also another type of traffic present, the network's own control traffic.

There are two major differences between control and customer traffic. The first is that control traffic results from the network operation and protocol signaling and provides the baseline for the connectivity between service end points on top of which customer traffic rides. In other words, control traffic is what keeps the network alive and breathing. The second difference is that the source of control traffic is internal. It is generated inside the network, while the customer traffic that transits the network between the service end points comes from outside the network.

As a traffic group, control traffic encompasses several and different types of traffic, for example, that for routing protocols and management sessions.

Let us provide a practical example illustrated in Figure 3.20. Here, customer traffic crosses the network, and router A issues a telnet packet destined to router B. Also because a routing protocol is running between the two routers, hello packets are present.

The two routers have different perspectives regarding the control packets. Router A has no ingress interface for them, because they are locally generated control packets. As for router B, the control packets arrive on a core-facing interface together with any other traffic that is being transmitted between the two routers.

Starting by looking at router A, how it deals with locally generated control traffic is highly vendor specific. Each vendor has its implementation regarding the priorities assigned to the traffic, the QOS markings of such packets, and which egress queues are used.

For router B, the classifier has to account for the presence of control traffic so that it can be identified and differentiated from the customer traffic. This is just like what happens with any traffic that belongs to a specific class of service.

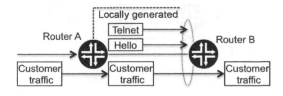

Figure 3.20   Control traffic in the network

It is good practice to keep the network control traffic isolated in a separate class of service and not to mix it with any other types of traffic due to its importance and unique character. Even if the behavioral requirements of control traffic are similar to those for other types of traffic, this is the only situation in which the similarities should be ignored and a separate class of service should be reserved for control traffic.

3.8   Trust, Granularity, and Control Traffic

We have been discussing several of the challenges that exist in a QOS network, including the trust borders between a network and its neighbors, the different granularity levels to be considered, and the presence of control traffic. Now we bring these three elements together and present a generic high-level view of a traffic flow across two routers.

In Figure 3.21, a unidirectional traffic stream (for ease of understanding) flows from customer site number 1 to site number 2. Packets belonging to this stream are represented as white packets, and the service is supported by two routers, named R1 and R2.

The first point encountered by the customer traffic flow is the service end point at the customer-facing interface on R1. This is the border between the customer and the service and, being a service end point, its granularity level is the highest so it should be customer aware.

Traffic originating from the customer is, in principle, considered untrustworthy, so the first step is to classify the customer traffic by assigning it to a class of service and to enforce any service agreements made with the customer, usually by applying the metering and policing functionalities or even more refined filtering based on other parameters that can go beyond the input rate of the traffic or its QOS markings.

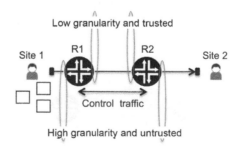

Figure 3.21   Traffic flow across two network routers

Although traffic can be assigned to multiple classes of service, we assume in this example that it is all assigned to the same class of service.

The next interface is the core-facing interface that connects routers 1 and 2, which the customer traffic is crossing in the outbound direction. As with any core-facing interface, it should be class-of-service aware, not customer aware, in terms of granularity. The customer traffic is grouped with all other traffic that belongs to the same class of service and queued and scheduled together with it. Because it is a core-facing interface, in addition to customer traffic, network control traffic is also present, so the queuing and scheduling rules need to account for it. Optionally at this point, rewrite rules can be applied, if necessary, to signal to the next downstream router any desired packets differentiation or to correct any incorrect markings present in the packets received from the customer. This particular step is illustrated in Figure 3.22. Triangular packets represent control traffic packets that exist on any core-facing interface. White and black packets are mixed in the same queue because they belong to the same class of service, with white packets belonging to this particular customer and black packets representing any other traffic that belongs to the same class of service.

Now moving to the core interface of router 2. The first step is classification; that is, inspecting the packets' QOS markings and deciding the class of service to which they should be assigned. Being a core-facing interface, the classifier also needs to account for the presence of network control traffic. All the traffic received is considered to be trusted, because it was sent from another router inside the network, so usually there is no need for metering and policing or any other similar mechanisms.

The final point to consider is the customer-facing interface, where traffic exits via the service end point toward the customer. Being a service end point, it is typically customer and class-of-service aware in terms of granularity. At this point, besides the queuing and scheduling, other QOS tools such as shaping can be applied, depending on the service characteristics. This example can be seen as the typical scenario but, as always, exceptions can exist.

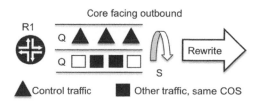

Figure 3.22   Queuing on a core-facing interface

3.9 Conclusion

Throughout this chapter, we have focused on the challenges that the reader will find in the vast majority of QOS deployments. The definition of classes of service is a crucial first step, identifying how many different behaviors are required in the network and avoiding an approach of "the more, the better."

Another key point that now starts to become visible is that a QOS deployment cannot be set up in isolation. Rather, it depends closely on other network parameters and processes. The network physical interfaces speeds and physical media used to interconnect the routers have inherent delay factors associated with them. Also, the network routing process may selectively allow different types of traffic to cross certain network hops, which translates into different traffic types being subject to different delay factors. So how the QOS deployment interacts with the existing network processes may limit or expand its potential results.

Further Reading

Davie, B., Charny, A., Bennett, J.C.R., Benson, K., Le Boudec, J.Y., Courtney, W., Davari, S., Firoiu, V. and Stiliadis, D. (2002) RFC 3246, An Expedited Forwarding PHB (Per-Hop Behavior), March 2002. https://tools.ietf.org/html/rfc3246 (accessed August 19, 2015).

4

Special Traffic Types and Networks

So far in this book, we have made a distinction between real-time and nonreal-time traffic as the major differentiating factor regarding the tolerance toward delay, jitter, and packet loss. We have also kept the traffic flows simple, just a series of Ethernet, IP, or MPLS packets crossing one or more devices.

This is clearly an over simplification, so in this chapter we take one step further by examining some of the more special scenarios. The environment where a QOS deployment is performed, either in terms of the type of traffic and/ or type of network present, always needs to be taken into consideration. However, some environments are more special than others due to their uniqueness or just because they are a novelty; as such the authors of this book have selected the following scenarios as the most "special ones" in their opinion, and each one will be detailed throughout this chapter:

- Layer 4 transport protocols—The User Datagram Protocol (UDP) and Transmission Control Protocol (TCP). Understanding how traffic is transported is crucial, for example, if the transport layer can retransmit traffic, is packet loss really an issue? Or how does the transport layer react when facing packet loss? How can it be optimal?
- Data Center—An area where the developments in recent years mandate a special attention. We will analyze storage traffic, the creation of lossless Ethernet networks, and the challenges posed by virtualization. Also there's

QOS-Enabled Networks: Tools and Foundations, Second Edition. Miguel Barreiros and Peter Lundqvist.
© 2016 John Wiley & Sons, Ltd. Published 2016 by John Wiley & Sons, Ltd.

the current industry buzzword Software-Defined Networks (SDN), is it a game changer in terms of QOS?

• Real-time traffic—We will analyze two different applications: a voice call and IPTV. The aim is to identify the underlying differences between different applications and secondly to decompose real-time traffic into its two components: signaling and the actual data.

While it should be noted that the above "special scenarios" can be combined, however, the aim of this book is to analyze each single one independently.

4.1   Layer 4 Transport Protocols: UDP and TCP

Let's start with UDP because it is the simplest one and it is a connectionless protocol; packets received by the destination are not acknowledged back to the sender, which is another way of saying that UDP is blind to congestion in the traffic path and if consequently packets are being dropped. The only assurance it delivers is packet integrity by using a checksum function.

So it is unreliable and it has no congestion control mechanism. However, it is extremely simple and highly popular.

The term unreliable should be seen in the context of the UDP itself; nothing stops the application layer at both ends of the traffic flow to talk to each other (at that layer) and signal that packets were dropped. For certain traffic types, such as real time, the time frame when the packet is received is crucial, so retransmission of traffic can be pointless.

There are several other characteristics of the UDP, but the key thing to retain is that the UDP is clueless regarding congestion and as such it has no adaptation skills.

The interesting Layer 4 transport protocol in terms of QOS is TCP, because it has two characteristics that make it special and popular, flow control and congestion avoidance tools, so as congestion happens, TCP changes its behavior.

It is not our goal to provide a detailed explanation of TCP; there are numerous papers and books that describe it in detail, and several of them are referenced in the further reading section of this chapter. Our goal is to present an oversimplified overview of TCP to arm the reader with the knowledge to understand its key components of slow start and congestion avoidance, which are commonly mentioned as the mechanisms for handling TCP congestion.

TCP is a connection-oriented protocol; it begins with a negotiation stage that leads to the creation of a session between sender and receiver. After the session is established when the receiver gets a packet from the sender, it issues an acknowledgement (ACK) back to the sender, which means the sender knows if the transmitted packet made it to the receiver or not. However, to avoid the sender waiting indefinitely for the ACK, there is a timeout value, after which the sender retransmits the packets. Also, TCP packets have a sequence number, for example, if the receiver gets packet number one and number three but not number two, it signals this event toward the sender by sending double ACK packets.

Now comes the adaptation part: TCP is robust, and it changes its behavior during the session duration according to the feedback that sender and receiver have regarding the traffic flow between them. For example, if packets are being loss, the sender will send those packets again and also it will send less packets next time. The "next time" is incredible vague on purpose, and we will exact the meaning further ahead, but as a teaser let us consider a theoretical and purely academic scenario illustrated in Figure 4.1.

Assume that for each packet that is lost the sender lowers the transmit rate by 50% and the network resource usage is at 90%, as illustrated in Figure 4.1. Although there are resources to serve packet number one, should it be dropped? If packet one is dropped, then "next time" the sender sends packet two and not two and three (50% transmit rate reduction). However, if packet one is not dropped, then "next time" both packets two and three are transmitted and then dropped due to a full queue, which forces the sender to stop transmitting packets at all (two times 50% reduction). Of course the reader may reverse the question, if there was only packet number one in Figure 4.1, why even consider to drop it? The answers are coming, but first we need to dive a bit further into TCP.

Back to our TCP session. What is transmitted between the sender and receiver is a stream of bytes, which is divided into segments, and each segment is send individually with a unique sequence number, as illustrated in Figure 4.2 for transferring a 3-megabyte file using a segment size of 1460 bytes.

This segmentation and numbering represents one of TCP's important features, since segments can arrive at the destination out of order, or sometimes a little

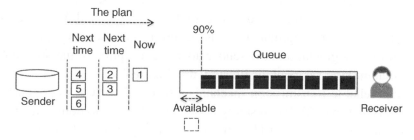

Figure 4.1 Proactive packet drop

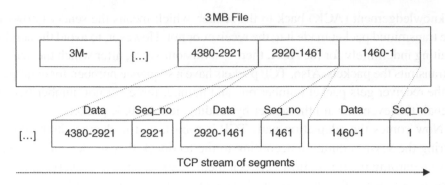

Figure 4.2   TCP segmentation

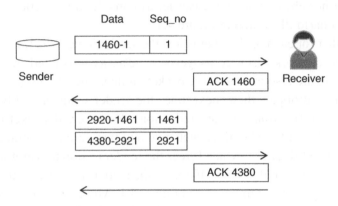

Figure 4.3   TCP acknowledge of segments

late, but it poses no problem because the destination can reorder it by keeping track of the byte positions associated with each segment's number.

Now comes another important feature; everything the destination receives is ACK back toward the sender, so the source knows what segments the destination has as illustrated in Figure 4.3.

Each segment, and thereby each received byte position, is ACK to the sender by the receiver. The ACK can be for one segment or be for several in one window of the transmission, as illustrated in Figure 4.3 for the second ACK packet. In the event there is one segment lost, the receiver starts sending double ACK packets for that segment, and the sender upon the reception of the double ACK resends that missing segment, as illustrated in Figure 4.4.

It is this reliable transmission mechanism that makes TCP so popular. The applications on top of TCP can safely trust its mechanism to deliver the traffic, so an application such as HTTP can focus on its specific tasks and leave the delivery of traffic to TCP, relying on it to fix any transport issues and to pack and

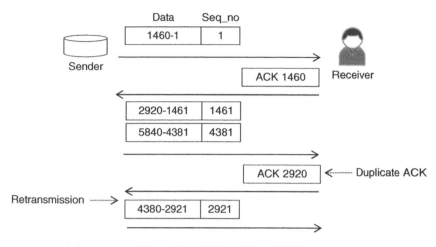

Figure 4.4   TCP duplicate ACK

unpack files in the proper order and byte position. All of this responsibility is left entirely to the end hosts (the sender and receiver), thereby relieving the intervening network transport elements from the need to perform transmission control. Just imagine if every router in the Internet were required to maintain state for each session, and be responsible, and ACK every packet on all links. For sure the Internet would not be what it is today if not for IP and TCP.

4.1.1   The TCP Session

Now let's focus on a real-life TCP session, a Secure Shell Copy (SCP) transfer (a random choice, SCP is just one to be one of many applications that uses TCP).

As illustrated in Figure 4.5, we have a sender (a server) with IP address 192.168.0.101 and a receiver (a host) with IP address 192.168.0.27, and just like any TCP session, it all starts with the classic three-way handshake with the sequence of packets SYN (synchronize), SYN-ACK, and ACK.

There are several parameters associated with a TCP session, but let's pick a few relevant ones for our current example.

The Maximum Segment Size (MSS) represents the largest chunk of data that TCP sends to the other end in a single segment. As illustrated in Figure 4.5, both end points of the session use 1460. As a side note, this value is typically inferred from the Maximum Transmission Unit (MTU) associated with each connection, so in the above example the connection is Ethernet (without VLAN tag) and the MTU is set to 1500 bytes; the MSS is equal to the MTU minus the TCP header (20 bytes) and the IP header (20 bytes)—that gives us 1460 bytes.

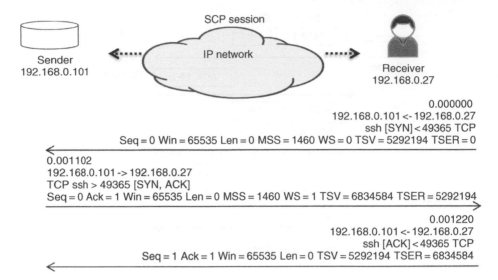

Figure 4.5   The TCP three-way handshake process

The window size (WIN) is the raw number of bytes that can be received or transmitted before sending or receiving an ACK. The receiver uses the receive window size to control the flow and rate from the sender, and a related important concept is the *sliding window*, which is the ability to change the window size over time to fully optimize the throughput.

The congestion window (CWND) defines what the sender can transmit while having outstanding ACK to be received, but for now suffice it to say that every time an ACK is received, it grows, and every time there is a timeout or the receiver flags that packets are missing, it diminishes.

Now that the TCP session is established, it is time to transfer real traffic for the first time. This is the stage named *TCP slow start*. Depending on the TCP stack implementation, the source will send one or two MSS full-sized packets; in this example let's assume two, as illustrated in Figure 4.6.

Because the packets arrive in sequence, the receiver needs to ACK only the last sequence byte size ending, which is 5798 (4350 + 1448), and this is how TCP optimizes the session speed. So all went smoothly and time has come to explain the "next time" concept mentioned briefly at the beginning of this section.

There is a successful transmission of data confirmed by the reception of the ACK packet, so the congestion window increases and next time we transmit more. This is what has earned the nickname of *greedy* to the TCP protocol; last time we transmitted two MSS full-sized packets, but now we will transmit nine MSS full-sized packets as illustrated in Figure 4.7.

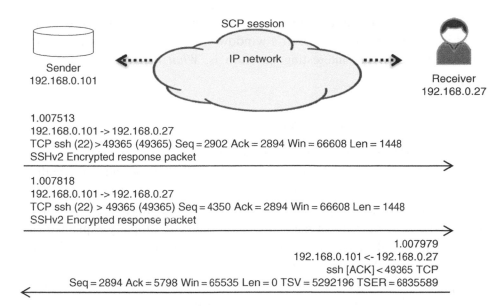

1.007513
192.168.0.101 -> 192.168.0.27
TCP ssh (22) > 49365 (49365) Seq = 2902 Ack = 2894 Win = 66608 Len = 1448
SSHv2 Encrypted response packet

1.007818
192.168.0.101 -> 192.168.0.27
TCP ssh (22) > 49365 (49365) Seq = 4350 Ack = 2894 Win = 66608 Len = 1448
SSHv2 Encrypted response packet

1.007979
192.168.0.101 <- 192.168.0.27
ssh [ACK] < 49365 TCP
Seq = 2894 Ack = 5798 Win = 65535 Len = 0 TSV = 5292196 TSER = 6835589

Figure 4.6   Double push

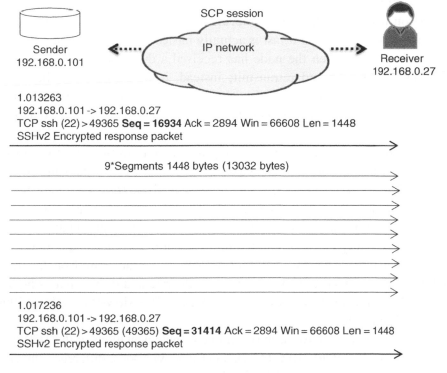

1.013263
192.168.0.101 -> 192.168.0.27
TCP ssh (22) > 49365 **Seq = 16934** Ack = 2894 Win = 66608 Len = 1448
SSHv2 Encrypted response packet

9*Segments 1448 bytes (13032 bytes)

1.017236
192.168.0.101 -> 192.168.0.27
TCP ssh (22) > 49365 (49365) **Seq = 31414** Ack = 2894 Win = 66608 Len = 1448
SSHv2 Encrypted response packet

Figure 4.7   Big TCP push

Where does it end? In a perfect world and a perfect network, the session's speed is ramped up to the receiver's window size and the sender's congestion window; however, the interesting question is, *What happens if packets are dropped?*

4.1.2   TCP Congestion Mechanism

The current TCP implementations in most operating systems are very much influenced by two legacy TCP implementations called *Tahoe* and *Reno*; both of them originate with UNIX BSD versions 4.3 and 4.4, and no doubt some eventful hiking trip. Their precise behavior (and of other more recent TCP stack implementations) in terms of mathematical formulas is properly explained and detailed in the references [1] at the end of this chapter, so here is just an oversimplified explanation of them.

So packet loss can be detected in two different manners: the timeout expired before the ACK packet was received, which always implies going back into the slow-start phase, or the sender is getting double ACK packets from the receiver.

For Tahoe, when three duplicate ACKs are received (four, if the first ACK has been counted), it performs a fast retransmission of the packet, sets the congestion window to the MSS value, and enters the slow-start state just like a timeout happened. Calling it fast is actually misleading; it's more of a delayed retransmission, so when the node has received a duplicate ACK, it does not immediately respond and retransmit; instead, it waits for more than three duplicate ACK before retransmitting the packet. The reason for this is to possibly save bandwidth and throughput in case the packet was reordered and not really dropped.

Reno is the same, but after performing the fast retransmission, it halves the congestion window (where Tahoe makes it equal to the MSS value) and enters a phase called *fast recovery*. Once an ACK is received from the client in this phase, Reno presumes the link is good again and negotiates a return to full speed. However, if there is a timeout, it returns to the slow-start phase.

The fast recovery phase is a huge improvement over Tahoe and comes as no surprise that most network operating systems today are based on the Reno implementation, because it tries to maintain the rate and throughput as high as it can and avoid as long as possible to fall back to TCP slow-start state, as illustrated in Figure 4.8.

In recent years there have been several new implementations of the TCP stack such as Vegas, New Reno (RFC 6582), and Cubic with some differences in terms of behavior. For example, the Cubic implementation that is the default in Linux at

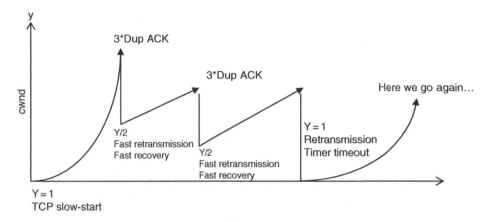

Figure 4.8 TCP throughput and congestion avoidance

the time of the writing of this book implements a cubic function, instead of linear, in the adjustment of the congestion window. The major novelty of the Vegas implementation is the usage of the round-trip time (RTT) parameter, the difference in time of sending a packet and receiving the corresponding ACK, which allows a more proactive behavior regarding when congestion starts to happen.

4.1.3   TCP Congestion Scenario

In the old slow networks based on modems and serial lines, retransmissions occurred right after the first duplicate ACK was received. It is not the same in high-performance networks with huge bandwidth pipes, where actual retransmission occurs some time after, because frankly, many packets are in flight. The issue gets more complex if a drop occurs and both the sender's congestion window and receiver's window are huge. The reordering of TCP segments becomes a reality, and the receiver needs to place the segments back into the right sequence without the session being dropped or else too many packets will need to be retransmitted.

The delay between the event that caused the loss, and the sender becoming aware of the loss and generating a retransmission, doesn't just depend on the time that the ACK packet takes from the receiver back to the sender. It also depends on the fact that many packets are "in flight" because the congestion windows on the sender and receiver can be large, so client retransmission sometimes occurs after a burst of packets from sender. The result in this example is a classic reorder scenario as illustrated in Figure 4.9.

In Figure 4.9, each duplicate ACK contains the SCE/SRE parameters, which are part of the Selective Acknowledgement (SACK) feature. At a high level,

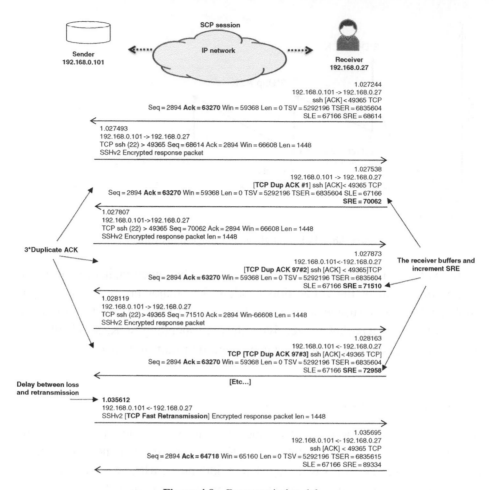

Figure 4.9   Retransmission delay

SACK allows the receiver to tell the sender (at the TCP protocol level) exactly which parts of a fragmented packet did not make it and ask for exactly those parts to be resent, rather than having the server send the whole packet again, which possibly could be in many fragments if the window is large.

In summary, TCP can take a fair beating and still be able to either maintain a high speed or ramp up the speed quickly again despite possible retransmissions.

4.1.4   TCP and QOS

The TCP throughput and congestion avoidance tools are tightly bound to each other when it comes to the performance of TCP sessions, and with current operating systems, a single drop or reordering of a segment does not cause too much harm to the sessions. The rate and size of both the receiver window and the sender congestion window are maintained very effectively and can be rapidly adjusted in response to duplicate ACK. However, there is a thin line between maintaining a good pace in the TCP sessions and avoiding too many retransmissions, which are a waste of network resources since there is no benefit in transmitting the same packet several times unless absolutely necessary.

There are several QOS tools discussed later in this book that are designed to handle the pace of TCP sessions, such as the queue dropper behavior, token bucket policing, or leaky bucket shaping.

Returning to the example of Figure 4.1, as the queue fill level approaches 100%, the dropper can start to drop "some" packets causing TCP to adapt instead of waiting for the queue to become full and then drop all packets forcing a return to the slow-start phase.

These tools are TCP friendly because they can be used to control whether a certain burst is allowed, thereby allowing the packet rate peak to exceed a certain bandwidth threshold to maintain predictable rates and stability, or to control bursts to stop issues associated with misconfigured or misbehaving sources as soon as possible without propagating that harmful traffic into the network. They can also implement large buffering to allow as much of burst as possible, thus avoiding retransmissions and being able to maintain the rate. Or in the other hand they can implement small buffering to avoid delay and jitter and also stop the transmit and receive window size becoming very large, which can result in massive amounts of retransmissions and reordering later on.

4.2 Data Center

Since the publication of the first edition of this book in 2010, the Data Center (DC) network evolution has been enormous.

It is now possible to create Ethernet networks with a lossless behavior, so that when facing congestion they have no packet loss, which, for instance, permits to transport Fiber Channel (FC) traffic as Fiber Channel over Ethernet (FCOE) among several other possible applications.

Another major change in the recent years is the massive deployment of virtualization; physical servers are not just physical servers anymore, but now they host multiple virtual machines (VMs) with a hypervisor as the front end.

And at last we now have Software-Defined Networks (SDN) allowing for faster and simpler deployment of connectivity between servers in the DC.

It is not the goal of this book to discuss the DC network design topics and their associated performance comparisons, for example, by comparing FCOE versus distributed file systems over IP, but solely to focus on the key points mentioned previously and the impacts in terms of QOS. But first let us describe storage traffic, because it is indeed a special traffic type.

4.2.1   SAN Traffic

Of all the traffic types that exist in a DC, the SAN traffic is special due to its requirements. If, in the process of writing information into a remote disk, packets are loss or placed out of order, then the disk is corrupted, so it has a zero tolerance regarding packet loss or packet reordering. And there is also the requirement for lower latency.

Networks can be either lossy or lossless, but the same principle applies to protocols; the TCP early described in this chapter assures a lossless environment by retransmitting packets that were lost, and by keeping track of the sequence numbers, it also assures that packet order is maintained. This is the basis of the Internet Small Computer System Interface (iSCSI) protocol, transmitting traffic over a lossy network with the TCP assuring the required lossless behavior.

As a side note, TCP is not an option for FCOE since FCOE is not IP based; however, let's keep comparing the different approaches to achieve lossless behavior.

There are two scenarios regarding at which OSI layer the lossless component is assured: at the transport layer with TCP or at a lower layer by the network itself. At first glance they might seem similar, but in reality they are different. The first scenario has already been detailed in this chapter with the TCP adaptation and congestion avoidance mechanisms. In the second scenario, the flow control and throughput reduction happen directly in the network devices. This can create concerns regarding how to apply flow control only to the flow causing congestion without affecting the other ones. It should be noted that with TCP packets are indeed lost, and then retransmitted, which is structurally different from not loosing packets in the first place.

There is also a third possible scenario, which is the combination of the preceding two. For example, iSCSI traffic can be transported over a lossless network; however, in that scenario the TCP adaptation and congestion mechanisms will never be used, because the network will perform flow control before TCP can act. This can potentially create issues, since TCP relies on its adaptation component to optimize the throughput.

But for now the most interesting scenario is how to transform an Ethernet network in a lossless one.

4.2.2 *Lossless Ethernet Networks*

In the FC networks the lossless component is assured via end-to-end signaling combined with a credit system, where the sender only sends traffic to the receiver if the receiver has buffers available, thus avoiding the situation of traffic arriving at the receiver and being dropped due to the lack of resources. This is flow control. There are a couple of drawbacks: FC devices are traditionally expensive, and running two separate networks always implies a higher cost and complexity from an operational and management point of view.

An Ethernet network doesn't have the credit capability like FC has; however, it has been given the PAUSE capability, which in a nutshell is the ability of a device to signal to its neighbor that it cannot transmit traffic and also its queue buffer is full, so it cannot store anything either. So to avoid dropping newly arrived packets, the neighbor needs to PAUSE the sending of traffic for a certain amount of time, as illustrated in Figure 4.10.

The PAUSE frame achieves the same result of the credit concept in the FC networks, flow control, controlling the flow of traffic along its path to assure that packets are not dropped. However, there are some drawbacks.

The generation and propagation of a PAUSE frame work in a hop-by-hop fashion and not from the congestion point directly toward the source of the traffic, so a congestion propagation phenomenon can happen. For example, in a traffic flow crossing a chained connection between devices one, two, and three, and when the queue buffer goes above a certain threshold in device three, device three signals that to device two, which then starts to store traffic in a queue instead of transmitting it to device three. However, that ability to store traffic is not infinite as previously detailed in Chapter 2, so if the congestion doesn't stop soon, then device two will also run out of resources and signal that to device one. So it can take a while until the original source of the traffic is paused, and

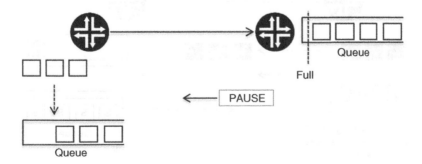

Figure 4.10   PAUSE frames

in the meantime all the devices along the traffic path are paused. This is commonly named the head-of-line (HOL) blocking scenario; if the congestion is not transient, the result could be that the flow of traffic is paused in all the devices it crosses, so the traffic flow is indeed controlled, but it is also completely paused.

To address this several protocols were created under the Data Center Bridging (DCB) umbrella. There are several DCB protocols and references to them at the end of this chapter, but the most interesting one from the perspective of this book is the Priority Flow Control (PFC).

An Ethernet frame has a QOS marking (a concept detailed in Chapter 5), and the generation of PAUSE frames will be relative to a specific QOS marking, effectively pausing just the traffic on the class of service that uses that QOS marking. So if SAN traffic uses a unique QOS marking, then PAUSE frames can be made specific for it, as illustrated in Figure 4.11 for three traffic types—black, gray, and SAN—each type with a different QOS marking.

The other relevant concept is the definition of priority groups, which is linked with the hierarchical scheduling concept that is detailed in Chapter 8. The goal is to apply queuing and scheduling first inside each priority group and then afterward apply queuing and scheduling to all the traffic, as illustrated in Figure 4.12.

So if SAN traffic has a unique QOS marking, then PAUSE frames are specific to SAN traffic, and also if SAN flows are assigned to a specific priority group, then they do not compete for the same resources with the other traffic types. Returning to Figure 4.12, such assurance can be given by the rightmost scheduler configuration, by, for example, giving the third queue a certain transmit rate. It should be noted that the lossless behavior is delivered by the existence of PAUSE frames, and what PFC allows is to leverage that functionalities even further.

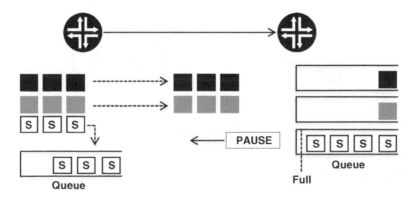

Figure 4.11   PAUSE frames per QOS marking

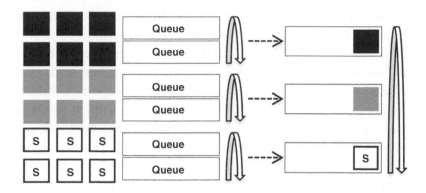

Figure 4.12   Priority Flow Control

4.2.3   Virtualization

The concept of virtualizing a physical host into multiple logical ones can be seen as the typical N:1 connection, where each VM talks to the hypervisor that then decides if the destination is another VM inside the same host or if the destination is outside this physical host. As with any N.1 connection, it is always vulnerable to congestion, and given the presence of different traffic types, then prioritization may also be required.

Today, hypervisors have mainly switching but also routing and security capabilities similar to any switch, router, or firewall, and the same applies for QOS. All the QOS tools discussed in this book are now starting to be available at the hypervisor allowing administrators to apply QOS directly at that level.

The real challenges in terms of QOS are the increase and the predictability of traffic flows between servers inside the DC (also commonly called the East–West traffic). This increase is due to a major change in how a "transaction" (in the lack of a better word) takes place between a user outside the DC and the resources inside the DC.

Let's use as an example a book purchase from an online store: the user opens its browser, does a search, and gets as a result a page with contents, pricing, pictures, and reviews among several other type of contents. Now, if all the book details and the web portal are in the same physical host in the DC, then this is simple in terms of traffic flows, because traffic enters the DC, arrives at a physical server where everything is stored, and the server responds to the user (ignoring gateways and firewalls for ease of understanding). This is typically called a North–South traffic flow. Well, this interaction belongs in the past because now all the book contents will be spread among different VMs, so the North–South traffic from the user to the DC will generate multiple West–East

communications between different VMs inside the DC to build all the contents
that the user sees in its browser, as illustrated in Figure 4.13.

Now comes the second challenge: today the book pictures could be stored in
a physical host on the West side and tomorrow in a physical host in the East side,
so how to predict the required resources inside the DC becomes a challenge.

The predictability factor is the hardest one to cope with. If two VMs are
communicating internally inside the physical host, then only the host hypervisor
deals with that traffic. However, when one of those VMs is moved to another
physical host, that same traffic that was previously "hidden" from the rest of the
network is now out there, using the DC devices and links that interconnect those
two physical hosts. And hundreds of VMs can be potentially moved across
physical hosts during a short period of time at just about any mid- to large-size
DC today. So this leads to the creation of traffic peaks that are not easily
predictable, and depending on the oversubscription ratios present in the DC,
buffering capabilities could be required. This is the scenario that is explored
beyond in the DC case study of this book.

Another scenario that has an impact is in the migration of a DC. When VMs
are being moved from the old to the new DC, there is traffic that before was only
seen by the host hypervisor that is now traveling across the two DC interconnection links demanding resources.

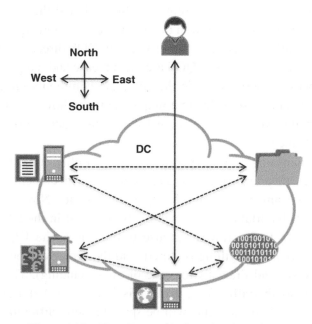

Figure 4.13   West–East traffic flows in a Data Center

4.2.4   Software Defined Networks

At the time of this writing, the industry's hot keyword is SDN, the ability to split the control and data planes in a network and deploy virtualized resources and build service chains with those virtualized (and physical) resources using orchestration tools and SDN controllers. The details around SDN are outside the scope of this book, but let's look at a typical chat between Viriatus (a server guy) and Viking (a networking guy) before SDN:

VIRIATUS	"I need connectivity between servers on racks 120 and 332 towards the server in rack 200, like a hub and spoke topology you know… Rack 200 is the hub"
VIKING	"What VLAN is that?"
VIRIATUS	"VLAN? I have no idea"
VIKING	"OK, no worries, I'll find out a VLAN number we can use, we'll actually need two, but leave it to me … but you need to be aware this is going to take a couple of days to set up"
VIRIATUS	"What? Why?"
VIKING	"Well I need to configure those VLANS on the top of the rack switches of racks 120, 332, and 200 an also configure Layer 3 interfaces to allow that inter-VLAN routing at our gateway router, and most important I need to make sure I don't break anything else, like using a VLAN number that is already being used"
VIRIATUS	"Damn …"
[AFTER 1 WEEK]	
VIKING	"Is it working now right?"
VIRIATUS	"No!"
VIKING	"Ah wait, must be the firewall, need to change that as well, sorry"

Now with SDN the story is different, and simpler, due to central controller nodes controlling the deployment of virtual networks and allowing to establish any form of desired connectivity between servers, without implying changes in the switching and routing infrastructure or by simply automating those changes to the underlay network. The goal is for the team that deploy and manage applications (and not the ones who own the infrastructure) to be able to request the network resources they need in a matter of seconds. This represents an obvious game changer. But how does that impact QOS?

If you look at SDN from the perspective of an automation process that allows application owners to create virtual networks, then if those virtual networks have QOS requirements or not become simply a part of such automation process. For example, if the application requires a VLAN to be created and used between three servers inside the DC, with or without QOS, the automation process will simply deploy QOS in that VLAN or not.

In terms of functionality there is nothing really new. QOS has proven itself in the past in both the nonoverlay (e.g., in Ethernet switching networks) and in the overlay (e.g., L2VPN or L3VPN) realms, so any SDN flavor is unlikely to impose dramatic changes on the way QOS is implemented.

However, for the SDN use cases where individual per-subscriber and per-application flows are being identified, the granularity with which QOS can be applied may become much tighter. This is due to a closer binding between the network devices where the traffic travels via the data plane and the applications, so features such as admission control on a per-application basis can be easily deployed. The same granularity can be applied regarding paths in the network and bandwidth traffic engineering.

4.2.5   DC and QOS

The ability to create Ethernet lossless networks sounds perfect at first glance; however, it does create some challenges such as congestion propagation and HOL blocking. For example, at present the design of a DC network regarding the transport of FCoE is typically made using a simple logical topology with a small number of hops, and then special care is given to the oversubscription ratios.

The major challenge created regarding virtualization is the increase and pre-dictability of East–West traffic. To cope with it, the first crucial step is visibility, typically achieved using telemetry and analytics. After visibility is present, a careful analysis of the oversubscription rations, and the deployment of buffering is typically required as well, which are topics further explored in the DC case study in the third part of this book.

SDN is a game changer in many ways, but in terms of QOS it is not so much; it is not expected to dramatically change the way QOS is applied, but it can allow for more granular deployments.

4.3   Real-Time Traffic

The handling of real-time traffic in packet-based forwarding networks is one of the major challenges Internet carriers, operators, and large enterprises have faced in the recent past, because it is structurally different from other traffic types. *How is it different?* First, because it is not adaptive in any way, it should come at no surprise that UDP is typically the transport protocol. There can be encoders and playback buffering enabled, but in its end-to-end delivery, it needs a constant care and feeding of resources. Second, it is selfish, but not greedy, and it requires specific traffic delivery parameters to achieve its quality levels, and ensuring those quality levels typically demands lots of network resources. If the traffic volume is always the same, the delivery does not change, but for that to happen, it needs protection and baby-sitting since rarely can lost frames be restored or recovered, and there is only one quality level: good quality.

There are several shapes and formats of real-time traffic, but the two most widely deployed are voice encapsulated into IP frames and streamed video encapsulated into IP frames, where typically short frames carry voice and larger frames carry streamed video. It can get more interesting as these services can be mixed in the same infrastructure, for example, the same 3G mobile phone receives data, voice, and video over the same infrastructure.

What all forms of real-time traffic have in common is the demand for predictable and consistent delivery and quality, and there is little or no room for variation in the networking service. The good news is that in most situations the bandwidth, delay, and jitter requirements can be calculated and therefore predicted.

4.3.1   Control and Data Traffic

For many forms of real-time applications, Real-Time Protocol (RTP) is the most suitable transport helper and is often called the worker of a real-time session, because it carries the actual real-time data, such as the voice or streamed media, where the signaling is delivered by the RTP Control Protocol (RTCP). But let's focus on RTP for a moment.

What RTP delivers is the transformation of IP packets into a stream and places four key parameters in the packet header: the sequence number, the timestamp, the type, and the synchronization source identifier (SSRC). The sequence number allows one to keep track of the order of the frames, enabling packet loss detection but also packet sequence restoration if frames are out of order. The timestamp reflects the sampling instant of the first octet in the RTP data packet—it is a clock that increments monotonically and linearly to allow synchronization and jitter calculations so that the original timing of the payload can be assembled on the egress side, thus helping maintain that payload quality. The Type field represents the type of data that is in the payload, and the SSRC identifies the originator of the packet.

As for the control part, the RTCP delivers the signaling for RTP, providing feedback on the quality of the data distribution and statistics about the session. Its packets are transmitted regularly within the stream of RTP packets, as illustrated in Figure 4.14.

Besides all the typical differences in the rates, as illustrated in Figure 4.14, the RTP packet size depends on its contents so they can be small or large while the RTCP packets are typically small.

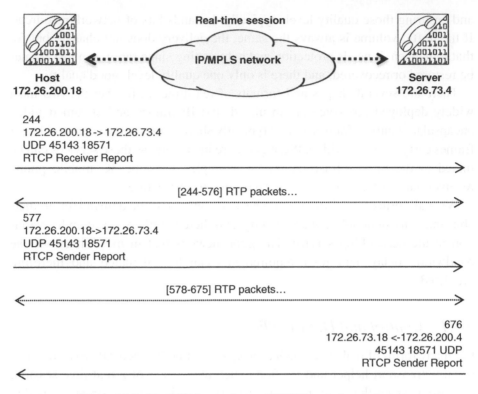

Figure 4.14   The RTP data packets and RCTP control packet rates

The point here is to highlight that real-time traffic is not just real-time traffic; there are a control and a data component.

Regardless of the difference in packet size and rates, RTP and RTCP packets belong to the same real-time session and are so closely bound together that in most scenarios it makes sense to place them in the same class of service. However, if RTP is being used to transport streaming video (large packets) and also voice (small packets), mixing them in the same queue can raise some challenges in terms of dimensioning the queue size and predicting the reaction to bursts of traffic, a topic that is explored further ahead.

4.3.2   Voice over IP

To further highlight the amount of different components that real-time traffic can have, let's examine the Voice over IP (VoIP) realm and the most popular IP-based voice protocol, Session Initiation Protocol (SIP). SIP is widely used for controlling multimedia communication sessions such as voice and video

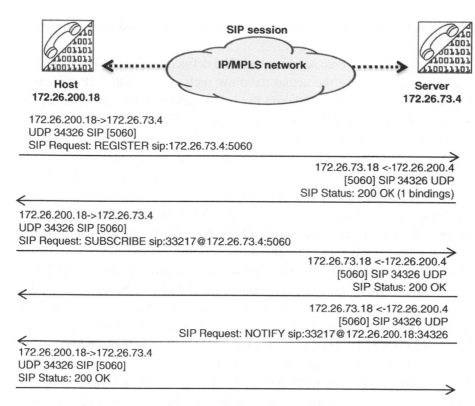

Figure 4.15   SIP register process, the classical 200 OK messages

calls over IP. It provides the signaling, and once a call is set up, the actual data (bearer) is carried by another protocol such as the RTP described previously.

In its most basic form, SIP is a context-based application very similar to HTTP, and it uses a request/response transaction model, where each transaction consists of a client request that invokes at least one response. Take, for example, Figure 4.15, where there is a soft phone registering with a server.

And there is more happening here. For example, typically Session Description Protocol (SDP) will select the codec used at egress among other items negotiated in the SIP call, but Domain Name System (DNS) will be used to perform name resolution and Network Time Protocol (NTP) will be used to set the time, and it should be noted that there is no transmitting of any real data traffic yet.

Because very different types of traffic are associated with SIP, placing all SIP-"related" packets in the same traffic class of service and treating them with the same QOS in the network are neither possible nor recommended. Imagine a DNS query sharing the same class of service of voice packets. Again, real-time traffic is not just real-time traffic; there are the control and data components associated with it.

So what part of the voice session should be defined as VoIP and be handled as such? Should it only be the actual dataflow such as the UDP RTP packets on already established calls? Or should the setup packets in the SIP control plane also be included? Should the service involved with DNS, NTP, and so forth also be part of the same class of service? Or are there actually two services that should be separated into different classes of service, one for the control and one for the forwarding? The dilemma here is that when everything is classified as top priority, then nothing is "special" anymore.

Delay and jitter for data packets can cause a lot of harm. Typically each device has some form of jitter buffer to handle arrival variations, and a well-known recommendation is that its value should be less than 30 ms.

The delay value is a bigger issue, and an old recommendation in ITU-TG114 suggests a maximum of a 150-ms end-to-end delay to achieve good quality, which may sound a bit too conservative, but note that this is end-to-end delay; thus in a network with several possible congestion points, any one of them can easily ruin the whole service. As for packet loss, it is actually better to drop something than to try to reorder it, because packets can remain in a jitter/playback buffer only for a limited time. If a packet is delayed for too long, it is of no use to the destination, so it is a mistake to spend network resources processing it.

Control packets typically cannot be dropped, but some delay and jitter variation is acceptable, because this signaling part of VoIP rarely affects any interaction with the end user other than the possible frustration of not being able to establish the call "right now."

The difference between control and data traffic is one of the catch-22 situations with VoIP. It is interesting to go back to basics and visit the PSTN realm and ask how did they do it. They calculate the service based on an estimate of the "average" load, from which they estimate the resources (time slots for bearer channels and out-of-band connections for the signaling). VoIP is not that different.

4.3.3 IPTV

Another highly popular application of real-time traffic is IPTV, the streaming of TV or video, which is essentially broadcasted over IP networks. Also note that streaming TV is often combined with voice in a triple-play package that delivers TV, telephony, and data to the end user subscriber, all over an IP network.

The focus here is strictly on QOS for IPTV delivery, and we will shy away from detailing the many digital TV formats (compressed and noncompressed, MPEG-2, and MPEG-4, SDTV and HDTV, and so forth).

The quality requirements for IPTV are very similar with those described earlier for VoIP, since both services need to have resources allocated for them and neither can be dropped. From the generation that grew up with analogue TV and the challenges of getting the antenna in the right position, there may be a little more tolerance for a limited number of dropped frames, whether this is a generic human acceptance or a generation gap is up to the reader. However, there are several fundamental differences between IPTV and VoIP. The most obvious one is packet size because IPTV packets are bigger and are subject to an additional burst, so IPTV requires larger buffers than VoIP. The delay requirement is also different, because the decoder box or node on the egress most often has a playback buffer that can be up to several seconds long, so jitter is generally not significant if a packet arrives within the acceptable time slot. Reordering of packets is possible, but it is not always trivial to do so, because each IPTV packet is large and the length of the playback buffer is translated into a maximum number of packets that can be stored, a number that is likely not to be large enough if a major reordering of packets is necessary. Other protocols involved in IPTV can be DHCP and HTTP or SSL/TLS traffic for images; however, these protocols for the most part do not visibly hurt the end user other than displaying a "Downloading… Please wait" screen message. The loss of frames seems worse to IPTV viewers than the dropping of words in a VoIP communication, but maybe because of the simple fact that most of the time, humans can ask each other to repeat the last sentence.

4.3.4   QOS and Real-Time Traffic

Real-time traffic requirements are often summarized as "minimum delay, minimum jitter, and no packet loss"; however, it always has two components, the control and the data. Whether it makes sense to map both of them in the same class of service depends on their requirements, but typically control traffic is more tolerant regarding delay and jitter. Also the control component of real-time traffic can contain several different protocols as highlighted in the VoIP call with the SIP.

Another key point is that real-time traffic presents itself in many different formats and typically with different requirements. The reasons why this book referred VoIP and IPTV is because they are popular and commonly used and secondly because although they seem similar at first glance, they have several significant differences such as their tolerance to jitter, which serves the purpose of illustrating that real-time traffic requirements depend on what is the traffic and how is being delivered.

Reference

[1] Marrone, L.A., Barbieri, A. and Robles, M. (2013) TCP Performance—CUBIC, Vegas & Reno, *Journal of Computer Science and Technology*, Vol. 13. http://journal.info.unlp.edu.ar/journal/journal35/papers/JCST-Apr13-1.pdf (accessed August 19, 2015).

Further Reading

Allman, M., Paxson, V. and Stevens, W. (1999) RFC 2581, TCP Congestion Control, April 1999. https://tools.ietf.org/rfc/rfc2581.txt (accessed August 21, 2015).

Chuck, S. (2002) Supporting Differentiated Service Classes: TCP Congestion Control Mechanisms, Whitepaper, Juniper Networks. www.juniper.net (accessed September 8, 2015).

DARPA (1981) RFC 793, Transmission Control Protocol—DARPA Internet Protocol Specification, September 1981. https://tools.ietf.org/rfc/rfc793.txt (accessed August 21, 2015).

Handley, M., Jacobson, V. and Perkins, C. (2006) RFC 4566, SDP: Session Description Protocol, July 2006. https://www.ietf.org/rfc/rfc4566.txt (accessed August 21, 2015).

IEEE, 802.1Qbb—Priority-based Flow Control. http://www.ieee802.org/1/pages/802.1bb.html (accessed August 21, 2015).

IEEE, 802.1Qau—Congestion Notification. http://www.ieee802.org/1/pages/802.1au.html (accessed August 21, 2015).

IEEE, 802.1Qaz—Enhanced Transmission Selection. http://www.ieee802.org/1/pages/802.1az.html (accessed August 21, 2015).

Mathis, M., Mahdavi, J., Floyd, S. and Romanow, A. (1996) RFC 2018, TCP Selective Acknowledgment Options, October 1996. https://www.rfc-editor.org/rfc/rfc2018.txt (accessed August 21, 2015).

Postel, J. (1980) RFC 768, User Datagram Protocol, August 1980. http://www.rfc-base.org/txt/rfc-768.txt (accessed August 21, 2015).

Rosenberg, J., Schulzrinne, H., Camarillo, G., Johnston, A., Peterson, J., Sparks, R., Handley, M. and Schooler, E. (2002) RFC 3261, SIP: Session Initiation Protocol, June 2002. https://www.ietf.org/rfc/rfc3261.txt (accessed August 21, 2015).

Schulzrinne, H. and Casner, S. (2003) RFC 3551, RTP Profile for Audio and Video Conferences with Minimal Control, July 2003. https://www.ietf.org/rfc/rfc3551.txt (accessed August 21, 2015).

Schulzrinne, H., Casner, S., Frederick, R. and Jacobson, V. (2003) RFC 3550, RTP: A Transport Protocol for Real-Time Applications, July 2003. https://tools.ietf.org/rfc/rfc3550.txt (accessed August 21, 2015).

Part II

Tools

Part II

Tools

5

Classifiers

In Part One of this book, we discussed the concept of per-hop behavior (PHB), in which each router in the network independently implements its own QOS settings because no signaling is possible either between neighbors or end to end. Consistency is achieved by applying a QOS configuration on each router in the network that applies the same PHB to traffic belonging to each class of service at each hop along the path that the traffic takes through the network.

In Part One, we also described the classifier tool, which assigns a class of service to each packet. This assignment is a crucial and fundamental step in the QOS realm because identifying traffic is the key factor in knowing the class of service to which it belongs. It is the class of service that determines the PHB that is applied to the traffic.

As previously discussed, the classifier tool has one input and N possible outputs. The input is the packet itself, and the N possible outputs are the number of different classes of service into which the packet can be classified. Classification can be considered to be an operation implemented using a set of IF/THEN rules, applied in the form of "IF (match conditions), THEN (class-of-service assignment)."

In this chapter, we focus on the types of classification processes that are currently commonly used, and we examine the differences between them in terms of the options available for the "IF (match conditions)" set of rules applied by the classifier. We also compare and debate the usage of the different classification processes.

QOS-Enabled Networks: Tools and Foundations, Second Edition. Miguel Barreiros and Peter Lundqvist.
© 2016 John Wiley & Sons, Ltd. Published 2016 by John Wiley & Sons, Ltd.

At the end of the chapter, we discuss the challenges of classifying traffic across multiple realms and also briefly introduce MPLS DiffServ-TE and its application regarding classification processes.

This book focuses on three specific realms, Ethernet, IP, and MPLS. In this chapter, when a specific realm is mentioned, the realm relates to the level at which the packet is processed by the classifier. For example, in a network in which IP (OSI Layer 3) packets travel over an Ethernet (OSI Layer 2) support, the network is called an Ethernet or an IP realm according to whether the classifier inspects the packet as an Ethernet or as an IP packet.

5.1 Packet QOS Markings

Currently, classifier processes based on the packets' QOS markings are the most straightforward and unquestionably the most popular ones. Each packet contains a field in which the QOS markings are located, although the exact location and length of the field vary according to the technology. However, the key point is that the standard for each technology defines a QOS field, so it is mandatory for that field to always be present.

The classifier inspects the packet, targeting the known location of the QOS marking, reads the bits in that field, and then makes a classification decision. Viewed from the perspective of the IF/THEN set of rules, the classifier takes the generic format of "IF packet QOS marking = X, THEN assign it to the class of service Y."

The packet classification process is straightforward because the classifier knows where in each packet to look for the QOS markings and because all packets always have those markings present at that location, since the location and the length of the packets' QOS markings are defined as part of the technology standards.

The only drawback of this type of classification process is that the number of bits used in the QOS field is necessarily finite for any given technology, so there is always a maximum number of different possible values available for a packet's QOS marking.

In the Ethernet realm, a VLAN-tagged Ethernet packet has a standard field called "User Priority" that is three bits long, thus providing up to eight different values for the QOS marking.

Similarly, in the MPLS realm, an MPLS label contains a standard field called experimental bits (EXP) that is also three bits long, providing up to eight different values for the packets QOS marking just as in the Ethernet realm.

In both Ethernet and MPLS realms, it is also possible to implement stacking, which consists of packets with multiple VLAN tags or MPLS labels, respectively.

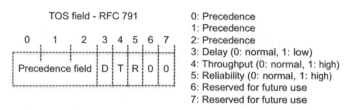

Figure 5.1   TOS field in IPv4 packets

Using as an example an Ethernet packet with two VLAN tags, the classifier can use as parameters the QOS marking in the outer VLAN tag or in the inner VLAN tag, or both. The exact same logic applies to MPLS packets with multiple labels.

In the IPv4 realm, things are slightly different. Inside each IP packet there is a standard field called type of service (TOS) that is eight bits long, as illustrated in Figure 5.1.

The first three bits (zero through two) are called the Precedence field, and it is in this field that the QOS markings are placed, thus providing eight different possible QOS markings. Bits three, four, and five define service requirements regarding delay, throughput, and reliability. However, these bits were never properly implemented and are not honored by the majority of the applications in the IP realm, so it's fair to say that only the first three bits in the TOS field are relevant.

RFC 2474 [1] redefined the TOS field in the IP header as a 6-bit Differentiated Services Code Point (DSCP) field, so bits zero through five are all available for QOS markings, and bits six and seven continue to be reserved. The DSCP field boosts the possible distinct QOS markings from 8 in the IP Precedence realm to 64 in the IP DSCP. From a conceptual point of view, there are no differences between using classifiers based on the IP Precedence or on the DSCP field; it's just a question of how many bits of the IP packet the classifier evaluates. However, care should be taken not to mix IP Precedence and DSCP classifiers, because bits three to five are irrelevant for IP Precedence but are relevant for DSCP. In the IPv6 realm, there is no such distinction: IPv6 packets contain a single field that is used for QOS markings.

5.2   Inbound Interface Information

Another parameter that can be used for classifying traffic is the interface on which the packet was received. Note that throughout this chapter, the term *interface* can mean a logical interface (such as an Ethernet VLAN or an MPLS tunnel) or a physical interface (e.g., a Gigabit Ethernet interface).

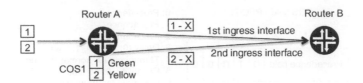

Figure 5.2   Classification based on the packets' QOS markings and input interface

Being able to use the inbound interface in a standalone fashion translates into very simple classification rules. Thus, all traffic received on a specific interface belongs to a specific class of service. From the perspective of the IF/THEN set of rules, the rules have the generic form "IF traffic is received on interface X, THEN it belongs to class of service Y."

There is an implicit trade-off between simplicity and flexibility, because such types of rules are inadequate for scenarios in which interfaces receive traffic that belongs to multiple classes of service. In fact, it is for this reason that this classification process is not commonly deployed on its own. However, when combined with other classification processes, knowing the inbound interface can improve the granularity of classification, for example, allowing a match on the packets' QOS markings and also on the interface on which they were received. Let us present an example of this approach by using the scenario in Figure 5.2, which shows two packets with the same QOS marking of X. We wish to signal to our downstream neighbor that each packet should be treated differently. Previously in Chapter 2, we solved this problem by using the rewrite tool on router A.

Router A has two packets numbered 1 and 2, and both of them are classified in the class of service named COS1. However, packet 2 is marked as yellow, a classification that we want to signal to the next downstream router.

The solution illustrated in Figure 5.2 is to have a classification process on router B that is based on the packets' QOS markings and also on the interfaces on which the packets are received. This means that a marking of X on the first ingress interface represents a packet that belongs to the class of service COS1 and that is green, while the same marking on the second ingress interface translates into the same class of service but with the color yellow.

So combining the processes for interface and QOS marking classification allows the same packet QOS marking to have a different meaning depending on the interface on which it is received.

However, the price for this is an increase in complexity, because this mechanism requires the use of multiple interfaces to connect the two routers (although the interfaces can simply be different logical interfaces on the same physical interface) and also because interface-specific classifiers must be defined.

In summary, interface-based classification is usually too rigid to be used in a standalone fashion, although when possible it is the simplest classification process possible. When combined with classification based on the packets' QOS markings, interface-based classification opens a pathway for packets with the same QOS marking to be assigned to different classes of service according to the interface on which they were received, a scenario that we explore later in this chapter.

5.3 Deep Packet Inspection

Deep packet inspection enters the next level of packet classification, in terms of depth, going beyond the straightforward packet QOS markings and moving from a macroscopic level to a microscopic one. The advantage of the microscopic level is granularity. For example, two packets with the exact same QOS marking can belong to different OSI Layer 4 protocols, or they can belong to the same Layer 4 protocol but have different session attributes, such as using different port numbers. Or going even further, discovering and identifying P2P (peer-to-peer) traffic on a network is a difficult task, because even at Layer 4 it is not possible to map this type of traffic to fixed or well-known port numbers. The only visible proof for P2P traffic is the traffic pattern and volume, and so only by inspecting the packet payload can P2P traffic be identified.

Deep packet inspection provides a huge boost in term of the match parameters that can be used in the IF/THEN set of rules implemented by a classifier. However, too much granularity can have a negative impact when it comes to the implementation stage. If the rules use highly granular parameters, then to cover all possible scenarios and all possible combination, it is necessary to have a huge number of rules. Let us give an example of this behavior for a classifier that matches on Layer 4 protocol and on the source port. If a classifier rule is defined to map TCP traffic with a source port of 80 to a certain class of service, it is also necessary to define rules for all other possible combinations of traffic that are not TCP and that do not use source port 80.

The situation illustrated in this example is the reason why deep packet inspection is usually not used as a standalone classification process. Its granularity can lead to the definition of a huge number of rules to cover all possible scenarios.

However, deep packet inspection can be combined with simpler classification processes to help pinpoint certain types of traffic that require a more in-depth inspection while leaving the handling of all possible scenarios to the simpler classification process.

5.4   Selecting Classifiers

At the time of writing, classification based on the packets' QOS markings is unquestionably the most popular and commonly deployed classification process. The main reason is ease of implementation, because each Ethernet, IP, and MPLS packet has a standardized field in which the QOS marking is present. This means that the classifier knows beforehand that a QOS field is present in the packet and knows its exact location.

As previously mentioned, the only drawback is that such fields have a finite length, so they have a maximum number of possible different values. However, in a certain sense, such a limitation has a positive implication in terms of simplicity and scaling, because the maximum number of rules that need to be defined (in the format "IF match marking, THEN …") is also limited.

The two other types of classifiers already discussed, interface-based and deep packet inspection, have their flaws as standalone processes. Interface-based classifiers are often too rigid, and deep packet inspection classifiers can drastically increase granularity but potentially leading to the creation of a large number of classification rules. However, these two classifiers can, without a doubt, become powerful allies for a simpler classification process, such as one based on the packets' QOS markings, by increasing its granularity or pinpointing specific types of traffic.

When selecting a classification process, one factor to take into account is the lack of signaling inside the network. One result of this limitation is that the only way to signal information between adjacent routers is to change the parameters used by the classifier on the next downstream router. If the classifiers are based on the packets' QOS markings, this task is achieved by using the rewrite tool, as explained in Part One. This points to another advantage of using the packets' QOS markings: because the QOS markings are in a standardized field, this field is present in all packets, and changing it is bound to have an effect on the classification process of the next downstream router. Interface-based classifiers obviously do not allow this flexibility, and deep packet inspection has the drawback that rewriting information present at Layer 4 or above can effectively impact the service to which the packet belongs.

For these reasons, this book always uses classification based on the packets' QOS markings by default, although scenarios of combining this classification process with others are also explored.

5.5 The QOS Network Perspective

From a network perspective, there are several possibilities regarding the classification process, but they all serve the same purpose: assigning packets to a certain class of service. A QOS-enabled network has two or more classes of service, because if there were only one, it would effectively be equivalent to a network without QOS.

When a generic packet named X with a certain QOS marking arrives at a router, the classifier inspects the packet's markings. Then, according to its set of IF/THEN rules, the router decides to which class of service the packet belongs, which translates into the PHB the router applies to it, as illustrated in Figure 5.3. There is no signaling, so this process is repeated on all the routers that the packet encounters as it crosses the network.

As packet X crosses the network routers, its QOS marking can be changed if any of the routers applies the rewrite tool, or the marking can be left untouched.

This mechanism is the basis of the PHB concept: a packet arrives at a router, and it is classified into a class of service and treated inside the router according to the PHB specified for that class of service. When the packet leaves the router, the previously performed classification is effectively lost and all the classification steps are repeated on the next downstream router.

Coherency is achieved by deploying classifiers on each of the routers crossed by the packets that ensure that packets requiring a certain PHB from the network are classified into the appropriate class of service.

As previously discussed, the classifier tool differentiates between different packets by assigning them to different classes of service, which implies that the maximum number of different classes of service that can exist in a network is limited by the parameters used by the classification process. Let us illustrate this by returning to the previous example and adding more detail. Assume an IP

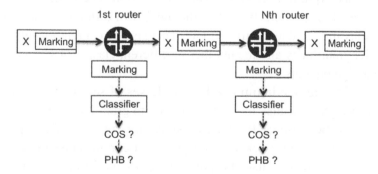

Figure 5.3 Classification occurs at each hop

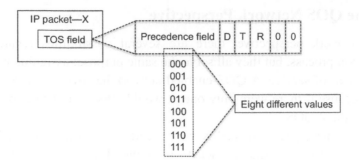

Figure 5.4   Possible values of the IP Precedence field

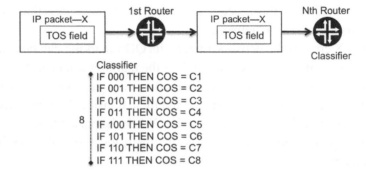

Figure 5.5   Classifier granularity and limit to the number of classes of service

realm in which the classifiers used by the routers are solely based on the IP Precedence field, which, as previously discussed, is a three-bit field, thus providing a maximum of eight different values, as illustrated in Figure 5.4.

Now, the router classifier looks at the IP Precedence field of the packets it receives and can find a maximum of eight different values. So from a classifier perspective, there are, at most, eight different types of packets, with each type identified by a unique IP Precedence bits combination. These combinations represent the classifier granularity, that is, the number of different packets between which the classifier is able to differentiate. There is a direct relationship between the classifier granularity and the number of different classes of service the classifier can assign packets to.

In Figure 5.5, the maximum number of classes of service the network can implement is eight, because this is the maximum granularity of the classifiers used. However, because of the lack of signaling, any specific requirements regarding differentiating packets belonging to a certain class of service that need to be signaled also use the packets' QOS markings, because this is the parameter used by the next downstream router. The result is a trade-off

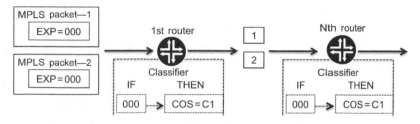

Figure 5.6   One EXP marking per class of service

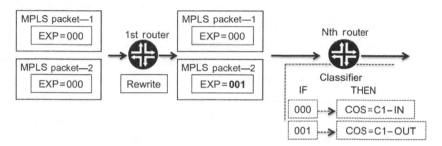

Figure 5.7   Two EXP markings per class of service

between the maximum number of classes of service and having more granularity within some of the classes of service.

Let us illustrate this behavior for an MPLS realm. We are purposely changing realms to illustrate that the PHB concept is indeed independent of the underlying technology. Figure 5.6 shows two MPLS packets crossing several network routers sequentially. The EXP field in both packets is zero, which means that they should be placed into the class of service C1, based on the set of rules present in the classifiers that inspect the EXP field.

Now assume that the first router makes a differentiation between the two packets, considering that both belong to the class of service C1 but that the first one is in contract and the second one is out of contract. If the first router desires to share this distinction with the next downstream router, it must change the parameter used by the next downstream router classifier, the EXP field, as illustrated in Figure 5.7.

By having the first router change the EXP value of the second packet to 001 and by properly deploying IF/THEN rules across the other network routers, it is possible to make the entire network aware that the first packet belongs to class of service C1 and is in-contract traffic, while the second packet belongs to the same class of service but represents out-of-contract traffic. Doing so effectively consumes two EXP values (out of eight maximum values) to provide differentiation within class of service C1, leaving six values available for use by other classes of service. Also bear in mind that all the possible EXP markings may not be available to differentiate between a customer's traffic because some of them may be required to identify network control traffic or other types of internal traffic.

In a nutshell, the classification process used by a network places limits both on how many different classes of service can exist in the network and on how information can be signaled to the next downstream router. So if the desire is to implement X number of classes of service in a network, the classification process chosen must be able to provide the necessary granularity while also allowing proper signaling between neighbors, if required.

Returning to the example of having eight different class-of-service values, eight may seem like a small number, but until now, both authors have not yet found a greenfield QOS network deployment that needed to implement more than eight classes of service. While this may sound odd, keep in mind that in a QOS world it's all about splitting resources across classes of service, so surely we can increase the divisor (number of classes of service), but the dividend (resources) remains constant.

As previously discussed in Chapter 3, the aim should always be to create the fewest possible number of classes of service.

5.6   MPLS DiffServ-TE

In Part One of this book, we highlighted the MPLS-TE capabilities of establishing LSPs with bandwidth reservations end to end and to have traffic follow a path different from the one selected by the routing protocol.

As a quick recap of what has been previously presented, bandwidth reservations limit the resource competition for bandwidth for traffic that is placed inside an LSP, because all other traffic present on the network does not compete for bandwidth access with traffic in the LSP. However, a gatekeeper mechanism at the ingress node is still required to enforce the fact that the bandwidth limit is not exceeded. Also, the ability to specify a path for the LSP different from the one chosen by the routing protocol creates the possibility of having multiple LSPs following different paths from each other, which affects the delay to which traffic inside each LSP is subject.

Although MPLS-TE has all these capabilities, it is blind in terms of the classes of service of the traffic it carries. An MPLS-TE LSP works simply as a tunnel that has the properties explained previously. MPLS DiffServ-TE LSPs (defined in RFC 3564 [2]) have all the previous properties and are also class of service aware.

The goal of MPLS DiffServ-TE is to allow the establishment of LSPs across the network, where each LSP contains traffic belonging to a specific class type (CT). A single CT can represent one or more classes of service, but for ease of understanding, let us assume a one-to-one relationship between CTs and classes of service. RFC 3564 defines eight CTs, although it is not mandatory to implement all eight across the network.

Across an MPLS DiffServ-TE network, each router has a traffic engineering database that specifies how much bandwidth is being consumed and how much is available to be reserved on a per-CT basis.

The classification processes at the ingress node, where the LSP starts, map traffic to a certain class of service based on a set of IF/THEN rules. Then, based on the assumption of a one-to-one relationship between CTs and classes of service, each MPLS DiffServ-TE LSP carries traffic belonging to one class of service. This assumption drastically simplifies the classification task for the transit nodes, because traffic received on one MPLS DiffServ-TE LSP belongs to only one class of service. Thus, in terms of the IF/THEN set of rules, the task on the transit nodes boils down to being able to infer the class of service to which the traffic belongs to, based on simply knowing the LSP on which the traffic was received. If we consider each MPLS DiffServ-TE LSP to be a logical interface, the dynamics are similar to an interface-based classification process, as illustrated in Figure 5.8.

The scenario in Figure 5.8 shows four types of packets. The classifier present on the ingress router classifies the packets into two classes of service, COS1 and COS2. The details of the classification process used by the ingress router are not relevant for this example. Now packets belonging to each class of service are placed inside a specific MPLS DiffServ-TE LSP, which greatly simplifies the task performed by the transit router: the class of service to which traffic belongs to is determined by knowing on which LSP the traffic was receive.

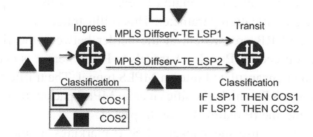

Figure 5.8   Classification in the MPLS DiffServ-TE realm

MPLS DiffServ-TE is a complex topic, and the previous paragraphs are an extremely simple introduction that focuses only on the classification processes. A reader not familiar with MPLS DiffServ-TE or seeking more detail should check out the further reading section at the end of this chapter.

5.7   Mixing Different QOS Realms

So far, we have discussed scenarios consisting of a single realm in which the same technology is used from the source to the destination. Single-realm networks make things easier from the perspective of classifying traffic. However, in the real world, it is common to have a mix of technologies between the traffic flow end points. For example, the end points can be located in IP realms that are connected by MPLS networks.

Conceptually speaking, from a traffic flow perspective, we can have a source network, one or more transit networks, and a destination network. The challenge with such scenarios is that when a change of realm results in a change in the granularity of the classifiers, lack of the number of classes of service and the associated PHBs that are available at each network may occur.

There are two distinct real-life scenarios. In the first, the middle network belongs to a service provider that offers a set of standard services from which customers choose to use to interconnect its sites or networks. In this scenario, there is not much room for discussion because the customer picks services from a fixed set of available options.

The second scenario is one in which the middle network belongs to the same administrative domain as the source and destination networks. Because all three network segments are in the same domain, the classes of service and associated PHBs that are available on the middle network can be tuned to achieve the desired end-to-end consistency.

Let us start with an example of two IP realms (called A and B) and an MPLS realm in the middle. Routers belonging to each domain use the packets QOS markings as the classification parameter. This is illustrated in Figure 5.9.

Routers RA and RB are present at the border between the IP and MPLS realms. Traffic sourced in IP realm A travels to RA, where it is classified as an IP packet and then encapsulated into an MPLS tunnel. The traffic travels inside this tunnel to router RB, which classifies it as an MPLS packet, then decapsulates it, and sends it into the IP realm B as an IP packet.

The two IP realms use the packets' DSCP markings as the classification parameter, which provides a granularity of up to 64 distinct QOS markings per

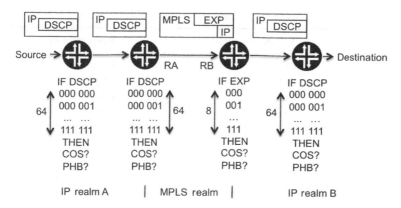

Figure 5.9   Connecting two IP networks via MPLS

packet. On the MPLS network (through which the IP traffic travels encapsulated), the classification parameter used is the EXP field, which provides up to eight different QOS markings. So this seems to be a problem, because two realms in which the classifier granularity is 64 are being connected by a realm in which the classifier granularity is eight.

In this situation, the first thing to bear in mind is that what matters is not the maximum granularity of the classifiers, which directly limits how many different classes of service can exist. The key point is not the maximum value but how many different classes of service must be implemented. That is, just because the DSCP field allows differentiating between 64 different types of packets, there is no obligation to implement 64 different classes of service in the network. The question should be how many classes of service are necessary rather than how to implement the maximum number.

Returning to the example in Figure 5.9, assume that IP realms A and B implement four different classes of service and packets belonging to each class of service are identified by their DSCP marking, as illustrated in Table 5.1.

The MPLS realm uses classification based on the EXP markings, which also allows a way to differentiate between four different types of packets. So consistency between the two realms is achieved by implementing four classes of service in the MPLS realm that are associated with the same PHBs as the four classes of service used in the IP realms, as illustrated in Table 5.2.

To make this example clearer, let us focus on a packet generated by the source that has the DSCP marking of 000 010. The packet arrives at router RA with this DSCP marking. The classifier inspects the packet's QOS field, finds the DSCP value, and based on the rules shown in Table 5.1, classifies it into the COS3 class of service. RA is the router sitting at the border between the IP A and MPLS realms,

Table 5.1   Classification rules in the IP realms A and B with four classes of service

IF DSCP field	THEN class of service	Respective PHB to be applied
000 000	COS1	PHB1
000 001	COS2	PHB2
000 010	COS3	PHB3
000 011	COS4	PHB4

Table 5.2   Classification rules in the MPLS realm

IF EXP field	THEN class of service	Respective PHB to be applied
000	COS1	PHB1
001	COS2	PHB2
010	COS3	PHB3
011	COS4	PHB4

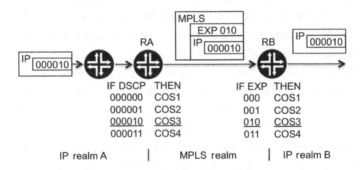

Figure 5.10   Packet traversing different realms

so it receives the packet as an IP packet but sends it toward RB with an MPLS label on top of it. So when the packet leaves RA toward RB, it is tagged by RA with an MPLS label that contains an EXP value of 010, as illustrated in Figure 5.10.

Router RB receives the packet with an EXP marking of 010, and following the rules in Table 5.2, it classifies it into the class of service COS3 and applies PHB3. Because RB is the router connecting the MPLS realm and the IP realm B, it removes the MPLS label before injecting the packet into IP realm B.

As long as on routers RA and RB the PHB applied to packets belonging to COS3 is the same (it should be PHB3, according to Tables 5.1 and 5.2), consistency is achieved from the source to the destination.

The integration illustrated in Figure 5.10 is straightforward, because the classifiers present in the MPLS realm have enough granularity to differentiate traffic into the four different classes of service that are implemented in the IP realms.

Table 5.3 Classification rules in the IP realms A and B with 10 classes of service

IF DSCP field	THEN class of service	Respective PHB to be applied
000 000	COS1	PHB1
000 001	COS2	PHB2
000 010	COS3	PHB3
000 011	COS4	PHB4
000 100	COS5	PHB5
000 101	COS6	PHB6
000 110	COS7	PHB7
000 111	COS8	PHB8
001 000	COS9	PHB9
001 001	COS10	PHB10

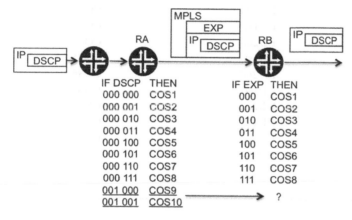

Figure 5.11 Exhaustion of the MPLS classifiers' granularity

The same scenario becomes more interesting if the IP realms implement more than 8 classes of services. Table 5.3 illustrates a case with 10 classes of service.

Having 10 classes of service exhausts the granularity of the classifiers present in the MPLS realm, as illustrated in Figure 5.11.

If an IP packet arrives at router RA with an DSCP marking of 001 000, it is classified into the class of service COS9. However, all the EXP markings inside the MPLS realm are already being used to identify traffic belonging to the classes of service from COS1 to COS8, so no more EXP values are available to identify traffic belonging to the class of service COS9, or, for that matter, to COS10.

As with many things in the QOS realm, there are several possible ways to solve this problem. We present one solution based on increasing the classifier granularity by using the information regarding the inbound interface on which traffic is received.

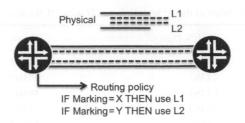

Figure 5.12 Mapping traffic to different tunnels according to the QOS marking

Table 5.4 Classification rules using the EXP field and interface

IF EXP	Tunnel/logical interface	THEN class of service
000	L1/1st	COS1
	L2/2nd	COS2
001	L1/1st	COS3
	L2/2nd	COS4
010	L1/1st	COS5
	L2/2nd	COS6
011	L1/1st	COS7
	L2/2nd	COS8
100	L1/1st	COS9
	L2/2nd	COS10
101	Unused	
110		
111		

The physical interface that connects routers RA and RB is split into two logical interfaces. Two MPLS tunnels (named L1 and L2) are created between RA and RB, and L1 is established across the first logical interface and L2 over the second. At RA, rules are implemented so that the selection of the tunnel to reach RB (either L1 or L2) is done according to the packet's QOS marking. This is accomplished with the help of routing policies, as illustrated in Figure 5.12.

The result of this split is that router RB receives traffic from two separate MPLS tunnels on two separate logical interfaces. When this tunnel arrangement is combined with classifiers that look at the packets' QOS marking and that are specific for each logical interface, the desired 10 classes of service inside the MPLS realm can be implemented, as illustrated in Table 5.4.

Table 5.4 shows an overlap in that the same EXP marking is used to identify packets belonging to different classes of service. However, there is no ambiguity,

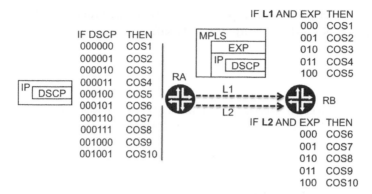

Figure 5.13   MPLS classifiers based on the EXP and tunnel from which traffic is received

because the value of the EXP field is used together with the information regarding the MPLS tunnel from which was the packet received. Returning to the network topology we used previously, Figure 5.13 illustrates the implementation of such a scheme.

This scheme does work and accomplishes the desired goal, but it is achieved at the expense of increasing the complexity of the network. The solution requires the creation of multiple tunnels between the source and the destination, the usage of logical interfaces, the introduction of routing policies, and the definition of classifiers on a per-interface basis. For a scenario with only two routers in the middle network, the task is easily doable. However, it is not quite the same with routers of the order of tens or hundreds.

5.8   Conclusion

Currently, classification based on packet QOS markings is without a doubt the most popular and widely deployed method, because of its simplicity in terms of using fields standardized for each technology and its ease of implementation. The other two types of classifiers presented in this chapter, interface-based and deep packet inspection, are commonly used as auxiliaries when it is necessary to increase granularity.

The maximum number of classes of service that can exist inside a QOS realm is limited by the classifier granularity. In addition, the required classifier granularity needs to be considered if there is a requirement to signal information between routers.

Mixing QOS realms may raise some challenges if there are differences between the granularity of the classifiers used in the different realms, as discussed in this chapter. The solution is typically to increase the granularity by combining the packets' QOS markings with other classification processes.

References

[1] Nichols, K., Blake, S., Baker, F. and Black, D. (1998) RFC2474, Definition of the Differentiated Services Field, December 1998. https://www.rfc-editor.org/rfc/rfc2474.txt (accessed September 8, 2015).

[2] Le Faucheur, F. and Lai, W. (2003) RFC3564, Requirements for Support of Differentiated Services-Aware MPLS Traffic Engineering, July 2003. https://www.ietf.org/rfc/rfc3564.txt (accessed August 19, 2015).

6

Policing and Shaping

The first part of this book presented the policer and shaper tools as black boxes and gave a high-level view of their functionalities. This chapter goes one step further and takes a closer look at the fundamental mechanics of these tools.

Policers are implemented using the concept of a token bucket, while shapers use leaky buckets. These are the two key concepts on which this chapter focuses. We explain both and compare the two in terms of applicability and differences regarding how to use them to deal with traffic burstiness and excess traffic.

6.1 Token Buckets

When viewed as a black box, the policer operation is pretty straightforward: at the ingress is a traffic flow with a certain bandwidth value and at the egress is the "same" traffic flow that is now enforced to conform to the bandwidth value defined by the policer. One result is that "excess traffic" is discarded, a behavior also commonly called rate limiting. This policer operation is illustrated in Figure 6.1.

In Figure 6.1, the policer limits the input traffic to a certain bandwidth limit (represented as a dotted line). During the time interval between t_0 and t_1, the input traffic exceeds the bandwidth limit value, which results in the excess traffic being discarded by the policer. While it is not mandatory, the usual action taken regarding excess traffic is to discard it. There are several other possible actions, such as accepting the excess traffic and marking it differently so that when both excess and nonexcess traffic types are past the policer tool, the router

QOS-Enabled Networks: Tools and Foundations, Second Edition. Miguel Barreiros and Peter Lundqvist.
© 2016 John Wiley & Sons, Ltd. Published 2016 by John Wiley & Sons, Ltd.

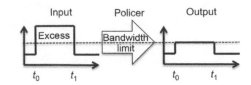

Figure 6.1   High-level view of the policing operation

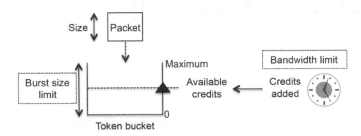

Figure 6.2   Token bucket structure

can still differentiate between them and apply different behaviors if desired. However, for ease of understanding, throughout this chapter we assume that the policer discards excess traffic, a behavior commonly called "hard policing."

The previous paragraph is a very quick recap of the high-level view of the policer tool as presented in Part One of this book. Now, let us look closer at the fundamental mechanics of the policer.

A policer is implemented using a token bucket algorithm for which there are two key parameters, the bandwidth limit and the burst size limit. The bandwidth limit is the rate at which traffic flows when leaving the policer, and the burst size limit parameter represents the policer's tolerance to traffic burstiness. These two parameters are usually measured in bits per second and bytes, respectively.

A token bucket operates using the concept of credits. When a packet of a certain size (also commonly called a length) arrives at the token bucket, the key question is whether there are enough credits to serve the packet. Put another way, the question is whether the packet should go through the policer (be transmitted) or should be discarded (be policed). The depth, or height, of the bucket is specified by the burst size limit parameter, while the bandwidth limit represents the credit refill rate, as illustrated in Figure 6.2.

So how does this credit scheme work? The triangle symbol in Figure 6.2 represents a meter that indicates how many credits are available inside the token bucket. As packets cross through the token bucket, the ones that are transmitted consume credits and the ones that are discarded do not. The burst size limit parameter specifies the bucket depth, that is, the maximum number of credits that are available at any given moment in the token bucket.

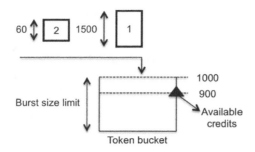

Figure 6.3 Two packets arriving at the token bucket

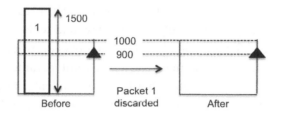

Figure 6.4 Packet discarded by the token bucket

The credit consumption of transmitting a packet is a function of its size, so every time a packet is transmitted, the triangle meter decreases as a function of the packet size. In parallel, the bucket is periodically refilled with a certain number of credits that is a function of the bandwidth limit parameter. But for now, let us pretend that the credits are not refilled.

As an example to illustrate how credits are consumed, let us start with a token bucket that has a burst size limit of 1000 bytes and that has 900 available credits. Two packets, numbered 1 and 2 with sizes of 1500 and 60 bytes, respectively, arrive sequentially at the token bucket, as illustrated in Figure 6.3.

The first packet, at 1500 bytes, is larger than the number of available credits, so it is discarded. The number of available credits in the token bucket remains at 900, because the packet was not transmitted, and hence no credits were consumed, as illustrated in Figure 6.4.

Before considering the second packet, it is worth pointing out that because the token bucket depth is 1000, any packets larger than this value are always discarded. Or put more generally, in a token bucket, packets whose size is larger than the burst size limit are always discarded.

The second packet is 60 bytes long, which is smaller than the number of available credits, so the packet is transmitted. The number of available credits becomes 840, which is the difference between the number available before the packet entered the bucket and the size of the transmitted packet, as illustrated in Figure 6.5.

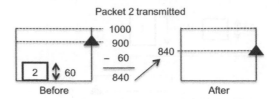

Figure 6.5   Packet transmitted by the token bucket

The reader can easily foresee from the logic of this example that the credits in the token bucket will always deplete completely. To refill the bucket, a mechanism of refilling the credits is working in parallel. The credit refill rate works in a timed fashion. Imagine that a clock is ticking, and with every single tick, a certain number of credits are added to the token bucket, where the number of credits added is a function of the bandwidth limit parameter and the time interval between each tick is usually a hardware-dependent parameter. However, credits are not added infinitely. As a rule, the maximum number of available credits that a token bucket can have at any time is always the value of the burst size limit value (the bucket depth). If no credits are being consumed because no packets are being transmitted, the clock continues to tick and credits continue to be added to the token bucket until the maximum value of available credits is reached and the bucket is full. This value is always capped at the burst size limit value.

We are now ready to present an illustration of the full operation of a token bucket that is simultaneously being depleted of credits and actively refilled with credits. The starting point is a token bucket configured with a burst size limit of 6000 bytes, and after the last packet was transmitted, the number of available credits is 1550 bytes. The bandwidth limit value is configured such that with every clock tick, 500 bytes worth of credits are added. For three consecutive clock ticks, no packets have arrived at the token bucket, thus raising the available credit value to 3050 (1550 plus three times 500). Now assume that after the end of the third, fourth, and fifth clock ticks, three packets numbered 1 through 3, all with a size of 2000 bytes, arrive at the token bucket, as illustrated in Figure 6.6.

In this example, we synchronize the arrival of the packets with the end of clock ticks. This is done solely for ease of understanding.

When packet 1 is placed in the token bucket, the number of credits available is 3050, which is greater than the packet size. Therefore, this packet is transmitted, and the available credit value is updated to the value it had before packet 1 arrived (3050) minus packet 1's size (2000). This results in 1050 credits being available, as illustrated in Figure 6.7.

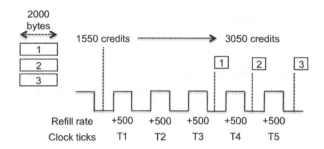

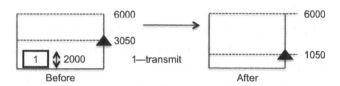

Figure 6.6 Starting point for a token bucket with credit refill

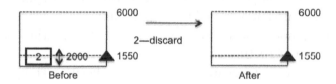

Figure 6.7 Packet 1 is transmitted

Figure 6.8 Packet 2 is discarded

After packet 1 is transmitted, clock tick T4 occurs, which adds 500 bytes of credits to the current value of 1050, thus resulting in a total of 1550 available credits. Now, following Figure 6.6, packet 2 arrives at the token bucket. The packet size of 2000 is larger than the number of available credits, so the packet is discarded and the number of credits remains at 1550, as illustrated in Figure 6.8.

At clock tick T5, 500 more bytes of credits are added to the token bucket, raising the number of available credits to 2050. The size of packet 3 is 2000 bytes, less than the number of credits available in the bucket, so this packet is transmitted, leaving the number of available credits at 50.

The aim of this example is to illustrate how a token bucket works with regard to the two values configured in a policer, the bandwidth limit and burst size limit. In a nutshell, the meter of available token bucket credits varies between zero and the burst size limit value, and the credit refill rate is a function of the configured bandwidth limit value.

As a teaser to open the topic of absorbing traffic bursts, the end result of the earlier example is that packets 1 and 3 are transmitted and packet 2 is discarded.

Now assume the same scenario but consider packets arriving at the token bucket at a faster pace. For example, if all three packets arrive between T3 and T4, only packet 1 is transmitted and the other two are dropped, because no credit refills occur until T4. As illustrated in Figure 6.7, after packet 1 is transmitted, the number of available credits is 1050, insufficient for transmitting either of the other two packets.

6.2   Traffic Bursts

A traffic burst can be seen as an abrupt variation in a traffic pattern. Bursts are an unavoidable fact in the networking world, because even if the traffic pattern rate is perfectly flat, just the initial jump or ramp from having no traffic whatsoever to that constant rate is itself a burst.

Let us start with an extreme example. Suppose a source of traffic is connected to a policer and the bandwidth limit implemented is 1000 bits per second (bps), which translates to 1 bit per millisecond. As illustrated in Figure 6.9, whether the source sends traffic at a rate of 1 bit per millisecond or sends 1000 bits in 1 ms, and nothing more is sent during that second, both scenarios conform to an average bandwidth of 1000 bps.

However, in the second scenario, there is a traffic burst, an abrupt variation in the traffic pattern, as illustrated in Figure 6.9, because all the traffic that could be said to be expected to be spread more or less across 1 s is sent in a single millisecond. What this means is a faster credit consumption in the token bucket in a shorter interval of time, which can lead to packets being discarded because as the time gap between the arrival of packet shrinks, fewer credit refill cycles occur, so a credit depletion scenario becomes more likely. As with many things in the QOS realm, discarding traffic bursts can be good or bad, depending on the desired goal.

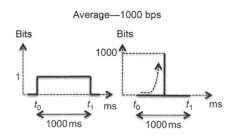

Figure 6.9   Traffic burst

Increasing the capability of the policer for burst absorption is directly connected to increasing the available credits present in the token bucket, which is a function of the burst size limit and the credits refill rate (which itself is a function of the bandwidth limit value). However, the bandwidth limit is a constant value, because it represents the bandwidth the policer must implement to handle the input traffic. If the desired goal of the policer is to implement an X bps rate, the bandwidth limit must be set to that value of X, which leaves only the burst size limit value as the only variable.

Another relevant factor is the size of the packets. In terms of credit consumption, ten 100-byte packets are equivalent to one 1000-byte packet, so the expected packet size should be taken into account when dimensioning the burst size limit of the policer. If the expected packet size is unknown, the only option is to consider the maximum packet size admissible into the logical or physical interface to which the policer is applied, a value called the maximum transmission unit (MTU).

Accepting traffic bursts needs to be considered not just from a policer perspective but also from a broader perspective as well, because the bursts place pressure on other QOS tools and on router resources. To illustrate this, we use the example illustrated in Figure 6.10. When traffic arrives at a router, it is classified and policed on the ingress interface, and as a result of the classification, it is mapped to queue A on the egress interface.

Assuming the rate purchased by the customer is X bps, this rate logically becomes the value to which the policer bandwidth limit parameter is set. In this scenario, the two remaining variables are the policer burst size limit and the length of queue A. Logically speaking, these two variables are interconnected. Accepting high traffic bursts at the policer makes no sense if the length of the output queue is small, because it will fill up quickly. The result is that packets that are part of that traffic burst are accepted by the policer, only to be dropped at the egress interface because the output queue has filled up. Thus these packets are consuming router resources in the transit path from the ingress interface until reaching the egress queue.

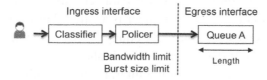

Figure 6.10   Ingress policer and output queue

As previously discussed, the queue length can be seen as a measure of the maximum possible delay introduced to a packet that is mapped to that queue. A longer queue allows more packets to be stored, at a possible cost of introducing more delay, while smaller queues minimize the risk of delay and jitter, at a cost of being able to store fewer packets.

The policer operation either transmits or discards a packet, but it never introduces any delay into the traffic transmission, so in the scenario illustrated in Figure 6.10, the delay inserted into the traffic crossing the router is controlled by the length of queue A. Once again, a division can be made between real-time and nonreal-time traffic. The burst size limit value should be smaller for real-time traffic than for nonreal-time traffic because queues used by real-time traffic ideally are small to minimize possible delay and jitter insertions. Keep in mind, however, that setting a very small burst size limit may lead to a scenario in which the credits in the token bucket are repeatedly being depleted.

For nonreal-time traffic, the burst size limit can be higher because the output queues usually also have a greater length, thus making it more likely for a high traffic burst admitted by the policer to fit in the output queue. However, this feature should not be seen as a criterion for using the highest possible value for the burst size limit parameter. It's quite the opposite, really. Returning to the extreme example illustrated in Figure 6.9, we have a 1000-bit burst in a single millisecond in a policer with a bandwidth limit value of 1000 bps. To absorb such a burst, all other traffic sent during the remaining 999 milliseconds of that second is discarded, because this is the only possible way for the policer to conform to the configured bandwidth limit value of 1000 bps. It is acceptable to absorb bursts, but not acceptable for a single traffic burst to consume the entire bandwidth limit. While it is true that this is an extreme example, the reader should always keep in mind that although some traffic burstiness can be accepted, the amount of burst tolerance should be kept within reasonable boundaries.

Unfortunately, there are no simple or bulletproof formulas to calculate the burst size limit. The usual approach taken in the field is to start with a small value and increase it using a trial-and-error approach that is generally based on the outcome of lab testing, real network experience, or recommendations in the router vendor's documentation. A starting point that both authors have found to be useful is to dimension the burst size parameter as a specific amount of the interface bandwidth on which the policer is applied. For example, on a gigabit interface, using a burst size limit value equal to 5 ms of the interface bandwidth results in a value of 625 000 bytes (1G multiplied by 5 ms and divided by 8 to convert from bits to bytes). When the interface bandwidth is considerably less than a gigabit, this rule may lead to a very small burst size value. In this case, another rule of thumb is to dimension the burst size limit in terms of how many packets are allowed in a burst, for example, 10 times the expected packet size or the link MTU. So, for example, if the MTU value configured is 1500 bytes, the burst size limit value is 15 000 bytes.

So far we have considered traffic that contains burst, abrupt variations in its pattern. To finalize the previous discussion, let us now consider that the traffic crosses the policer at a perfect flat rate, without any bursts, and also that the traffic rate is always below the policer bandwidth limit parameter. The key point to retain is that packets crossing the policer still consume credits, and as such, the burst size limit still needs to be adjusted to the expected pace at which packets arrive at the token bucket.

6.3 Dual-Rate Token Buckets

A policer can be defined to simply limit all traffic indiscriminately to a certain bandwidth, or it can be defined to be more granular. For an example of a scenario requiring greater granularity, consider that all traffic arriving from a customer should be globally policed to 10 Mbps and that the input traffic flow contains two types of traffic, voice and nonvoice. Voice traffic should be policed to 2 Mbps, and this rate must be guaranteed. That is, as long as voice traffic remains below the 2-Mbps barrier, it must always be transmitted, as illustrated in Figure 6.11.

The first possible solution to comply with these requirements is to use two independent token buckets to limit voice traffic to 2 Mbps and nonvoice traffic to 8 Mbps. This scheme guarantees that voice traffic never exceeds the 2-Mbps barrier and also meets the requirement for a guaranteed rate. However, if no voice traffic is present, the total amount of bandwidth that nonvoice traffic can use is nevertheless limited to 8 Mbps. The leftover bandwidth created by the absence of voice traffic is not accessible to nonvoice traffic, which can be good or bad depending on the desired behavior. This waste of bandwidth is the price

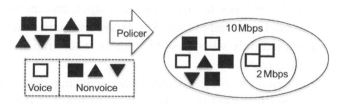

Figure 6.11   Different types of traffic with different policing requirements

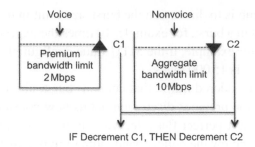

IF Decrement C1, THEN Decrement C2

Figure 6.12   Interconnection between the credit rates of two token buckets

to pay for placing traffic into independent token buckets. The lack of communication between the two buckets implies that the bucket into which nonvoice traffic is placed must implement the 8-Mbps bandwidth limit to assure that voice traffic always has access to its 2-Mbps guaranteed rate.

Another possible approach is defining two token buckets but linking the credit rates of both. In this approach, voice traffic is placed in a bucket called "premium" that imposes a 2-Mbps rate, and nonvoice traffic is placed in a bucket called "aggregate" that imposes a 10-Mbps rate. This scheme allows nonvoice traffic to use up to 10 Mbps in the absence of voice traffic. However, it raises the concern of how to assure the 2-Mbps guaranteed rate to voice traffic. The answer is to link the credit rates of both token buckets. Every time a voice packet is transmitted, the available credit meters of both the premium and the aggregate are decremented, while the transmission of a nonvoice packet decrements only the aggregate available credit meter, as illustrated in Figure 6.12.

Dual-rate policers are popular and are commonly implemented together with the metering tool. The most common implementation deployed is the two-rate, three-color marker, defined in RFC 2698 [1].

6.4   Shapers and Leaky Buckets

As previously discussed in Part One, the goal of a shaper is to make the received traffic rate conform to the bandwidth value in the shaper definition, also called the shaping rate. A shaper has one input and one output, and the output traffic flow conforms to the defined shaper rate. Any excess traffic is stored inside the shaper and is transmitted only when possible, that is, when transmitting it does not violate the shaping rate.

A shaper is implemented using a leaky bucket concept that consists of a queue of a certain length, called the delay buffer. A guard at the queue head assures that the rate of the traffic leaving the leaky bucket conforms to the

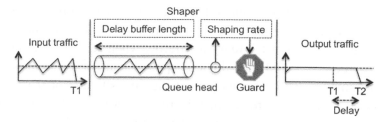

Figure 6.13 Leaky bucket operation

shaping rate value, as illustrated in Figure 6.13, which represents the shaping rate as a dotted line.

Usually, the shaping rate is measured in bits per second and the delay buffer length in milliseconds or microseconds. The shaping rate is a "constant" value in the sense that if the desired result is to have the input traffic shaped to X Mbps, the shaping rate should be set to the value X. The result is that only the delay buffer length is a variable.

Two factors must be taken into account when dimensioning the delay buffer length parameter. The first is that the length of the delay buffer is a finite value, so there is a maximum amount of excess traffic that can be stored before the delay buffer fills up and traffic starts being dropped. The second factor is that when excess traffic is placed inside the delay buffer, it is effectively being delayed, because the shaper guard enforces that traffic waits inside the delay buffer until transmitting it does not violate the shaper rate. This behavior is illustrated in Figure 6.13, which shows that the input traffic graph ends at T1, but the output traffic ends at T2. This is a key point to remember: the shaper's capability to store excess traffic is achieved at the expense of introducing delay into the transmission of traffic.

The delay buffer is effectively a queue, so quantifying how much delay is inserted is dependent on the queue length, its fill level (how many packets are already inside the queue), and the removal from the queue speed. The speed of removal from the queue is obviously a function of the shaping rate, which is itself set to a constant value—the desired rate for traffic exiting the shaper tool.

Predicting the queue fill level at a specific point in time is impossible. However, it is possible to analyze the worst-case scenario. When a packet is the last one to enter a full queue, the time it has to wait until it reaches the queue head (and thus to become the next packet to be removed from the queue) can be represented by the queue length value. Thus, the worst-case scenario of the delay introduced by the shaping tool can be represented as the length of the delay buffer. Following this logic for the jitter parameter, the least possible delay introduced is zero and the maximum is the length of the delay buffer, so the smaller the gap between these two values, the smaller the possible jitter insertion.

As previously discussed for real-time traffic, when sensitivity to delay and jitter is high, the shaper tool should not be applied at all, or it should be applied with a very small delay buffer length. At the opposite end of the spectrum, nonreal-time traffic has a greater margin for dimensioning the length of the delay buffer because this traffic is less sensitive to delay and jitter.

6.5   Excess Traffic and Oversubscription

The existence of excess traffic is commonly associated with the oversubscription phenomenon, the concept of having a logical interface with a certain bandwidth value that during some periods of time may receive more traffic than it can cope with, as illustrated in Figure 6.14.

Oversubscription is popular in scenarios in which it is fair to assume that all sources of traffic do not constantly transmit to the same destination simultaneously, so the bandwidth of the interface to the destination can be lower than the sum of the maximum rates that each source can transmit. However, with oversubscription, transient situations can occur in which the bandwidth of the interface to the destination is insufficient to cope with the amount of bandwidth being transmitted to it, as illustrated in Figure 6.14.

Oversubscription is a scenario commonly used with the shaping tool, because its application makes it possible to guarantee that the rate of traffic arriving at the logical interface to a destination complies with the shaper's rate. So by setting the shaper rate equal to the logical interface bandwidth, any excess traffic generated by the transient conditions is stored inside the shaper tool instead of being dropped. Such logic is valid for a shaping tool applied at either an ingress or egress interface.

A typical network topology that creates transient conditions of excess traffic is a hub and spoke, in which multiple spoke sites communicate with each other through a central hub site. In terms of the connectivity from the spoke sites

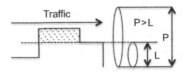

Figure 6.14   Oversubscription scenario. P, physical interface bandwidth; L, logical interface bandwidth

toward the hub, the bandwidth of the interface connecting to the hub site can be dimensioned as the sum of the maximum rates of each spoke site, or alternatively the bandwidth can be set at a lower value, and the shaping tool can absorb any excess traffic that may exist. However, the second approach works only if the presence of excess traffic is indeed a transient condition, meaning that the delay buffer of the shaper does not fill up and drop traffic, and only if the traffic being transmitted can cope with the delay that the shaping tool inserts.

This topology is just one scenario that illustrates the need for dealing with excess traffic. Several others exist, but the key concept in all scenarios is the same.

The business driver for oversubscription is cost savings, because it requires that less bandwidth be contracted from the network.

6.6   Comparing and Applying Policer and Shaper Tools

So far, this chapter has focused on presenting separately the details about the token and leaky bucket concepts used to implement the policer and shaper tools, respectively. This section compares them both side by side.

As a starting point, consider Table 6.1, which highlights the main characteristics of the two tools. The striking similarity is that the end goal of shapers and policers is the same: to have traffic that exits the tool conform to a certain bandwidth value expressed in bits per second. Let us now focus on the differences in terms of how the two achieve this same goal.

The first difference is that the policer introduces no delay. As previously discussed, the delay introduced by shaping is the time that packets have to wait inside the delay buffer until being transmitted, where, in the worst-case scenario, the delay introduced equals the length of the delay buffer. Obviously, delay is introduced only if there is excess traffic present. The fact that variations in the delay inserted are possible implies that the introduction of jitter is also a factor that must be accounted for. It is quite common in the networking world to use the analogy that the policer burst size limit parameter is a measure of delay

Table 6.1   Differences between policing and shaping

Tool	Input parameters	End goal	Delay introduced	Excess traffic
Policer	Bandwidth limit Burst size limit	Traffic conforming to a certain bandwidth value	No	Dropped
Shaper	Shaping rate Delay buffer length		Yes	Stored

introduced, but this is incorrect. The policer action introduces no delay whatsoever, independent of the burst size limit value configured.

The second difference is that the policer drops excess traffic (which is why no delay is inserted), but the shaper stores it while space is available in the delay buffer, dropping the excess traffic only after the buffer fills up. However, as previously discussed, the policer can absorb traffic bursts as a function of its burst size limit parameter. It is interesting to highlight the differences between absorbing traffic bursts and allowing what we are calling excess traffic.

Suppose that traffic is crossing the policer at a constant rate and that at a certain point in time the rate increases (i.e., bursts occur), thus increasing the arrival rate of packets at the token bucket. If the absorption of burst leads to a complete credit depletion, when the "next" packet arrives at the token bucket, no resources (credits) are available to transmit it, and it is dropped. Now, if we look at this situation from a shaper perspective, when a packet cannot be transmitted, it is not dropped, but rather it is stored inside the delay buffer. The packet waits until it can be transmitted without violating the shaping rate. The only possible exception is when the delay buffer is already full, and then the packet is dropped.

We mentioned at the beginning of this chapter that we are considering a policer action of discarding traffic that violates the defined policer rate (i.e., hard policing). Let us now examine the effects of using an action other than discard, an action commonly named soft policing. Consider a policer named X, whose action regarding traffic that exceeds the defined policer rate is to accept it and mark it differently (e.g., as "yellow," while traffic below the policer rate is marked as "green"). One could argue that policer X also accepts excess traffic, which is true. However, policer X is allowing yellow traffic to pass through it (not discarding it and not storing it but just marking it with a different color). So the total amount of traffic (green and yellow) present at the output of the policer X does not conform to the defined policer rate, because excess traffic is not discarded or stored.

Deciding where to apply the policer and shaper tools in the network effectively boils down to the specific behavior that is desired at each point. Taking into account the differences highlighted in Table 6.1 and in the previous paragraphs, the most striking difference between the two is how each one deals with excess traffic. Practical examples for the application of both tools are presented in the case studies in Part Three, but at this stage, we highlight some of the common application scenarios.

Some common scenarios involve a point in the network at which the bandwidth needs to be enforced at a certain value, either because of a commercial agreement, an oversubscribed interface, or a throughput reduction because of a

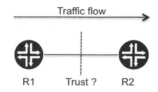

Figure 6.15   Scenario for policing and shaping applicability

mismatch between the amount of traffic one side can send and the other side can receive.

To analyze the applicability of the policing and shaping tools, let us use the scenario illustrated in Figure 6.15, which shows two routers R1 and R2, with traffic flowing from R1 toward R2.

From a router perspective, the policer and shaper tools can be applied to both ingress traffic entering the router and egress traffic leaving the router. The decision about where to apply the tools depends on the behavior that needs to be implemented, which is affected in part by the trust relationship between the two routers. For example, if R1 and R2 have an agreement regarding the maximum bandwidth that should be used on the link that interconnects the two routers, should R2 implement a tool to enforce the limit or can it trust R1 not to use more bandwidth than what has been agreed?

The trust parameter is a crucial one, and typically the network edge points (where the other side is considered untrustworthy) are where these types of tools are commonly applied.

Let us start by analyzing the scenario of excess traffic. If R1 has excess traffic, it has two options. The first is shape the traffic and then transmit it to R2. The second option is for R1 to transmit the traffic and for R2 to use ingress shaping to deal with the excess. If R1 considers R2 to be untrustworthy regarding how it deals with excess traffic, the second option should be ruled out. If, on the other hand, R1 considers R2 to be trusted, the shaping tool can be applied at the ingress on R2.

Regarding the need to enforce the traffic from R1 to R2 at a certain rate by using the policer tool, if the relationship in place between the two routers is one of trust, a tool to enforce the rate can be applied at the egress on R1 or at the ingress on R2. Ideally, though, the tool should be applied on the egress at R1 to avoid using link capacity for traffic that will just be dropped at the ingress on R2. If R2 does not trust R1 to limit the traffic it sends, the only option is for R2 to implement an ingress tool to enforce the limit.

The discussion in this section provides only generic guidelines. What the reader should focus on is the requirements to be implemented together with the capabilities and impacts of each tool. And you should never shy away from using the tools in a different way from the ones illustrated here.

6.7   Conclusion

Throughout this chapter, we have discussed the mechanics used to implement policing and shaping and the differences between them. The aim of both the policer and shaper tools is to impose an egress rate on the traffic that crosses it. The policer also imposes a limit on how bursty the traffic can be, while the shaper eliminates such traffic bursts at the expense of delaying traffic.

The burst size limit parameters represent the policer tolerance to traffic burstiness. However, even if the traffic is a constant flat rate below the policer bandwidth limit value and even if no bursts are present, this parameter needs to be dimensioned, because packets crossing the policer always consume credits.

The use of both tools should always be bound to the application requirements because, as highlighted throughout this chapter and also in Chapter 2, the end results can be very different.

Reference

[1]   Heinanen, J., Finland, T. and Gueri, R. (1999) RFC 2698, A Two-Rate Three Color Marker, September 1999. https://tools.ietf.org/html/rfc2698 (accessed August 18, 2015).

7

Queuing and Scheduling

Throughout this book, we have always referred to the queuing and scheduling tool as the star of the QOS realm, the tool that makes crystal clear the principle of benefiting some at the expense of others. In this chapter, we analyze the internals of the queuing and scheduling mechanism that allow such differentiation. But first, we analyze the parameters associated with any queuing and scheduling mechanism and their common ground. Only then do we present the different queuing and scheduling mechanisms. The most commonly deployed mechanism is PB-DWRR, which stands for priority-based deficit weighted round robin. However, we also present all other mechanisms that have preceded PB-DWRR because analyzing their pros and cons provides the necessary clues to understand the roots and goals of PB-DWRR and why the industry has moved toward it.

7.1 Queuing and Scheduling Concepts

Queuing and scheduling applied to an interface allows traffic to be split into multiple queues so that the scheduler can decide which type of treatment the traffic inside each queue receives. If the traffic mapped to each queue belongs to a specific class of service, the scheduler can apply differentiated behavior to different classes of service, as illustrated in Figure 7.1.

The above is a quick recap of what has been previously discussed throughout this book, so let us now dive more deeply into the internals of queuing and scheduling.

QOS-Enabled Networks: Tools and Foundations, Second Edition. Miguel Barreiros and Peter Lundqvist.
© 2016 John Wiley & Sons, Ltd. Published 2016 by John Wiley & Sons, Ltd.

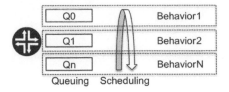

Figure 7.1 Queuing and scheduling applying different behaviors

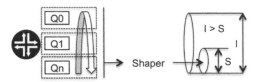

Figure 7.2 Bandwidth parameter in queuing and scheduling. I, interface bandwidth; S, shaping rate

The two most important parameters associated with the queuing and scheduling mechanism are buffers and bandwidth. Buffering is the length of the queue, that is, how much memory is available to store packets. However, the entire packet does not need to be queued, as we see later; sometimes what is stored is a *notification*, which is a representation of the packet contents. The buffering value can be defined either as a time value during which packets are accepted on the interface on which queuing is enabled or as a physical size in terms of how many packets or packet notifications can reside in the queue at the same time. The buffering value is a quota of available memory and can be defined as milliseconds of traffic or absolute numbers of packets.

The bandwidth parameter refers to the scheduling part of the equation. A total amount of bandwidth is made available to the queuing and scheduling mechanism. Scheduling determines how much is allocated to each queue. The total amount of bandwidth can be either the interface speed or the shaping rate if a shaper is applied after the scheduler, as illustrated in Figure 7.2.

The queuing and scheduling discipline used determines how the resources are allocated. The requirement for the presence of queuing and scheduling is typically controlled by the presence of congestion. If resources are sufficient and there is no competition for resources, there is no need for queuing. One way to create congestion is to place more traffic on an interface than the outgoing line speed can support. Congestion can also be created artificially by applying a shaping rate to the interface that imposes a speed limit lower than the maximum interface line speed. The leftover traffic is throttled or back pressured into memory, which is then partitioned across the actual queues. The scheduler then

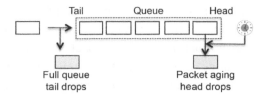

Figure 7.3   Tail and head aging drops in a queue

services the queues and is responsible for the rate at which packets from each queue are transmitted.

As previously discussed in Chapter 2, a queue has a head and a tail. A packet enters a queue at the tail, remains in the queue until it reaches the head, and then leaves the queue. In a queuing system, packets can be dropped from either the tail or the head of the queue and can even be dropped from both at the same time. Most commonly, packets are dropped from the tail. When the queue buffer fill rate is much faster than the removal rate, the buffer fills up completely. The result is that no more packets can be placed in the buffer and any newly arrived packet needs to be dropped. But a queue can also drop packets from the head. Packets at the head of the queue are the ones that have moved from the tail to the head and thus are those that have stayed in the queue for the longest amount of time. In the case of extreme congestion and resource starvation, the queue receives no scheduling slots. To avoid stalling the queue and having a hopelessly long delay for the traffic inside the queue, a maximum time is enforced on all packets in a queue for how long they are allowed to remain in the queue, waiting to be scheduled. The term for this is *packet aging*, meaning that the packets are aged out from the queue buffer because there is no point in trying to deliver them.

Tail drops and packet aging are not mutually exclusive. If, to drain the queue, the rate of dropping packets at the head of the queue cannot keep up with the rate at which packets are entering the queue at the tail, packet dropping can occur from both the tail and the head as a result of packet aging, as shown in Figure 7.3.

7.2   Packets and Cellification

Routers can handle packet queuing in two different ways, either queuing the entire packet or splitting it into fixed-size cells, a process commonly called cellification. Queuing the entire packet is the older method and is conceptually simpler. It is commonly used in CPU processing-based forwarding routers. In this scenario, there is no split between the control and forwarding planes, which typically does not become a limiting factor if the supported interfaces are low-speed ones. However, queuing the entire packet has its drawbacks, which have led to the appearance and popularity of cellification. A packet has a variable length, where the minimum and maximum values are specified by the technology or optionally by the interface configuration. Also, a packet can contain several different headers. These possible variations in the packet length and in the number of possible headers create the challenge of how to manage memory resources. Also, the router CPU is affected, because every time the packet is processed, it needs to be read bit by bit. So different packet lengths imply different processing delays. In Chapter 3, we discussed this topic briefly when we analyzed several different types of delay factors inherent to packet transmission. As we stated in Chapter 3, we ignore processing delay, which can be done if the control and forwarding planes of the router are independent.

The cellification process chops a packet into fixed-size cells, which makes it easier for the router's memory management to deal with the packets. It also provides consistency in terms of transmission times and slots for the buffering required. The cellification process is shown in Figure 7.4, which shows a packet containing several headers that is split into 64-byte cells.

Cellification also has a special cell called the notification, also called the "cookie." The cookie contains the relevant pieces of the packet that are required for its processing inside the router. So instead of processing the entire packet every single time, it needs to be evaluated by any tool, the router evaluates only the cookie. The idea is to create a representation of the packet with only the required information, for example, if an IP packet contains in its header the DSCP value of X that is contained in the cookie. However, other information, for example, the packet payload, may not be added to the cookie. The cookie contents are dynamic, so if the packet is assigned to a certain class of service, this information is written to the cookie. Also, if information contained in the packet header, such as the DSCP field, is changed, the cookie is updated as well. The cellification process is typically completely transparent both in terms of the

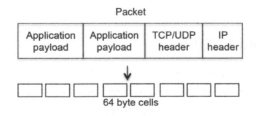

Figure 7.4   Cellification of a packet

router's configuration and management. Nevertheless, it is important to understand its basics and the benefits it offers.

7.3   Different Types of Queuing Disciplines

Over a number of years, several mathematical studies have led to the design and creation of a variety of queuing algorithms. To describe all possible queuing mechanisms is impossible. However, some major well-known disciplines regarding scheduling are detailed in the next sections of this chapter:

- First in, first out (FIFO) queuing
- Fair queuing (FQ)
- Priority queuing (PQ)
- Weighted fair queuing (WFQ)
- Weighted round robin (WRR)
- Deficit weighted round robin (DWRR)
- Priority-based deficit weighted round robin (PB-DWRR).

The most powerful of these and perhaps the most interesting is PB-DWRR. However, this chapter discusses them all, because it is interesting to evaluate the pros and cons of each one and because doing so illuminates the evolutionary history of queuing algorithms and the reasons behind the popularity of PB-DWRR.

7.4   FIFO

FIFO is a well-known acronym for first in, first out and is probably the most basic queuing scheduling discipline. The principle behind FIFO queuing is that all packets are treated equally by all being placed in the same queue. So in a nutshell, there is one queue and the scheduler only serves this queue. This mechanism implies that the removal rate is directly inherited from either the interface speed or from the shaping rate if there is a shaper applied. Because with any queuing mechanism there is no overtaking within a queue, the packets are serviced in the same order in which they were placed into the queue. This is illustrated in Figure 7.5.

For most router vendors, FIFO is probably the default behavior, with one queue for most transit traffic plus another one for locally generated control plane packets such as routing protocol packets.

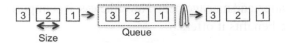

Figure 7.5   FIFO scheduling

Generally speaking, most hardware implementations need at least one buffer per interface to be able to build a packet before transmitting it. The function of the queue is to be a placeholder buffer that is used when the Layer 1 and Layer 2 headers are added to the packet. Providing a buffer that is the size of two to four packets helps the system to be able to utilize the interface efficiently and at line rate. FIFO service is predictable and the algorithm is simple to implement for a router or a switch. However, the FIFO algorithm is perhaps not a true "queuing and scheduling" solution because it does not provide any form of differentiated services, since traffic belonging to different classes of service share the same road lane (effectively the same queue) to the scheduler.

Under congestion conditions, FIFO queuing first introduces a delay as the queue starts to fill, and when the queue becomes full, all newly arrived packets are discarded. Here the implementation of the drop behavior can differentiate.

One alternative is to drop everything in the buffer to avoid having the queue collapse under the burden of stalled sessions. The reason for this behavior is to avoid a situation similar to the TCP silly syndrome, in which the receiving window gets smaller and smaller because most of the packets in the buffer memory are waiting to be processed. The result is that all sessions suffer service degradation and the packets are stuck in the queue in the order in which they arrived. Despite the possible availability of large buffer memory, the buffers cannot be utilized properly because only a small number of senders' congestion window (CWND) can be used.

Another drop alternative is to use the more intelligent ways described earlier to clear the queue by dropping from the tail and by using packet aging.

FIFO queuing has the following benefits:

- The FIFO algorithm is simple to implement.
- It provides acceptable performance if the queue saturation is low and short. In fact, FIFO can be seen as more of a burst-absorbing implementation than a true queuing strategy.
- If applications are TCP based and not delay sensitive, the process of removal from the queue is well suited because packet ordering is preserved at the cost of introducing delay. If moderate congestion occurs, TCP slows down because of RTT, but retransmissions are kept to a minimum.

FIFO queuing has the following limitations:

- FIFO provides no differentiated services. In cases of extreme congestion and buffer usage, all services are equally bad.
- Delay and jitter cannot be controlled because the queue depth usage can vary. Therefore, FIFO is not an appropriate solution for real-time applications. For example, a voice flow consisting of many packets might be stuck behind a 1500-byte TCP transfer with a large CWND, a concept discussed in Chapter 4.
- Greedy flows can take up most of the queue depth, and thus bursty flows can consume all available buffer space. Because TCP, by nature, tries to scale up the size of the transmitting window, session control is very difficult with a single queue or buffer.

7.5   FQ

The main disadvantage of FIFO queuing is that flows consisting of many packets can take up most of the bandwidth for less bandwidth-intensive applications, because FIFO does not separate flows or streams of packets. FQ, also commonly called the fairness algorithm, is a scheduling algorithm that addresses the basic limitation of FIFO queuing. FQ classifies packet flows into multiple queues, offering a fair scheduling scheme for the flows to access the link. In this way, FQ separates traffic and flows and avoids applications that consume less bandwidth being starved by applications that consume more bandwidth.

The scheduling is very much a statistical multiplexing process among the queues, with queues buffering packets that belong to different flows. FQ, defined by John Nagle in 1985, takes into account packet size to ensure that each flow receives an equal opportunity to transmit an equal amount of data. Each queue is assigned the same weight; that is, the queues are scheduled with the same amount of bandwidth. This scheme avoids the problems of dealing with small and large packets in the queue, so the speed of removal from the queue is a function of a number of bits, not a function of a number of packets. So, for example, a queue with large packets has access to the same bandwidth as a queue with small packets, because in each scheduling turn the queue is serviced in terms of number of bits, not number of packets.

The main issue with the fairness algorithm is the fact that it is indeed fair, which may seem odd, but it does invalidate the desired QOS goal of providing an uneven distribution of resources across different classes of service. If many queues must be visited in a fair order, flows that require low delay can suffer.

For routers to implement the FQ model, a hash function needs to separate flows from each other to map packets to a particular session. The hash function can compute the session or flow using a rich set of variables, such as a combination of the source port address, destination port addresses, and protocols, and possibly even higher-layer information beyond TCP, UDP, and port numbers. But once this classification is done, dynamically allocating memory and creating logical queues that exist only when the flow is active is a very resource-intensive task that is hard to implement when most of the packet forwarding takes place within a hardware-based architecture that has little interaction with processing resources. As a result, classifying and dynamically creating queues for each active hashed flow is an unscalable process.

The fairness algorithm is a common tactic to handle oversubscribed backplanes on routers and switches. A common hardware architecture is to have a one-stage to two-stage fabric switch that connects line modules. In this scenario, incoming traffic from multiple ports is forwarded over a fabric to a single destination port. Here, fairness can ensure that the incoming ports on the line modules have equal access to the fabric so that they can be delivered to the ports on the outgoing module, as illustrated in Figure 7.6.

However, the fairness algorithm does not take into account the fact that the backplane is not aware of individual flows, which are the basis for the original idea of using FQ to create bandwidth fairness. So what is indeed implemented is a slight variation of the original FQ algorithm.

The FQ algorithm implies that the data rate may fluctuate for short intervals, because the packets are delivered sequentially to achieve fairness in each scheduling cycle. This behavior can possibly lower the throughput, especially when compared to algorithms that focus more on maximum throughput. On the plus side, however, FQ is very effective in avoiding starvation of flows under heavy loads.

FQ offers the following benefit:

• The FQ algorithm isolates each flow into its own logical queue. Thus, in theory, a greedy flow cannot affect other queues.

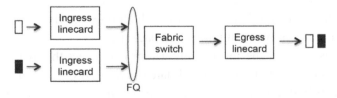

Figure 7.6   Fairness algorithm

FQ has the following limitations:

- The FQ algorithm is extremely complicated to implement, and no example of a vendor implementation exists on high-speed routers or routers created to handle large numbers of sessions and large amounts of traffic. FQ is more of a theoretical construct than a practical paradigm.
- It is resource intensive because many states and hashes must be computed and because memory must be allocated and reallocated based on changes in the session state.
- Delay and jitter can still be issues because each session hash is seen as a queue entity. If many sessions need to be scheduled, the multiplexing scheme used to determine the session slot interval can vary, depending on how many active sessions are actively transmitting packets. The wait time to the next scheduled service can be long if many sessions are active.

7.6   PQ

For applications that are sensitive to delay and that are not able to handle packet loss, the PQ scheduling algorithm provides a simple method of supporting differentiated service classes. After a classification scheme has classified the packets and placed them into different queues with different priorities, the PQ algorithm handles the scheduling of the packets following a priority-based model. Packets are scheduled to be transmitted from the head of a given queue only if all queues of higher priority are empty. So the different levels of priority assigned to queues introduce the unfairness desired regarding of how queues are serviced. However, inside a single priority queue, packets are still scheduled in a FIFO order, as with any other queuing and scheduling method. The PQ algorithm operation is illustrated in Figure 7.7.

In Figure 7.7, queue zero (Q0) has the highest priority, so as long as there are packets in Q0, the scheduler serves only this queue. Q1 has a low priority, so packets are removed from this queue only when Q0 is empty.

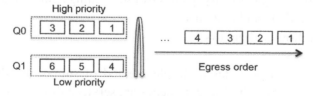

Figure 7.7   PQ scheduling

PQ offers benefits for traffic that requires no packet loss and low delay. With PQ, such applications can be selectively scheduled, and their service can be differentiated from the bulk of the best-effort flows. The PQ requirements for queuing and scheduling are minimal and not very complex compared with other more elaborate queuing disciplines that also offer differentiated service.

However, PQ has its own set of limitations. If the volume of high-priority traffic becomes excessive, lower-priority queues may face a complete resource starvation. If their queuing rate remains constant but the rate of removal from the queue decreases, packet drops start to occur, either from the tail or the head of the queue or possibly both. The drop rate for traffic placed in low-priority queues increases as the buffer space allocated to the low-priority queues starts to overflow. The result is not only packet drops but also increased latency. Regarding TCP sessions, the state of TCP traffic may become stale if it has to be retransmitted. Also, introducing such a conservative queuing discipline and scheduling mechanism can affect the network service to such extent that the service for the whole network decreases because of starvation of the majority of traffic and applications. As discussed in Chapter 4, some real-time applications require several packet types to achieve a real-time service. For example, for Session Initiation Protocol (SIP) traffic, the PQ algorithm can prioritize SIP traffic. However, if DHCP offers cannot reach users and DNS servers cannot resolve queries to reach SIP servers and users, the gains obtained by favoring SIP traffic are limited because no users are available to establish SIP sessions. In addition, misbehaving high-priority flows can add delay and jitter to other high-priority flows that share the same queue. Effectively, a real-time service comprises several different applications and packet types, and all these applications and packet types must receive equivalent service levels for the real-time service to operate properly. Having all one's eggs in the same basket is not a design for today's Internet and for large-scale enterprise networks.

PQ can be implemented both in a strict mode and in a rate-controlled mode. For example, the priority queue rate can be policed to drop or reclassify traffic if the rate increases beyond certain thresholds. Such thresholds can be specified as a percentage of the link bandwidth or an explicit rate. This method avoids starvation of lower-priority queues and can also provide control over the delay inserted in traffic in high-priority queues by establishing a limit on the maximum amount of traffic that can be queued.

PQ has the following benefits:

- The PQ algorithm provides a simple method of supporting differentiated service classes, in which different queues are serviced differently according to the priority assigned. The buffering and scheduling computational requirements

are low and not very complex, even when implementing PQ on network equipment with high-speed links.
- Low-delay and loss-sensitive applications such as real-time traffic can be effectively protected from other greedy traffic flows such as TCP sessions, and low delay and jitter can be maintained for the traffic classified as high priority.

PQ has the following limitations:

- Because of its absolute allocation of resource and scheduling services to the priority-classified traffic, the PQ algorithm can result in network malfunctioning because low-priority traffic becomes stalled. The cost of giving certain traffic a higher priority comes at the expense of penalizing lower-priority traffic.
- High-priority traffic must be very well provisioned and rate controlled. Otherwise, service breakdowns can result for all traffic that is not high priority.

7.7 WFQ

WFQ is commonly referred as "bit-by-bit round robin," because it implements a queuing and scheduling mechanism in which the queue servicing is based on bits instead of packets. WFQ was developed in 1989 by Demers, Keshav, Shenke, and Zhang and emulates the Generic Processor Sharing (GPS) concepts of virtual processing time. In WFQ, each queue or flow is allocated a weight that is a proportion of the interface rate or the shaping rate. WFQ is aware of packet sizes and can thus support variable-sized packets. The benefit is that sessions with big packets do not get more scheduling time than sessions with smaller packets, because effectively the focus of WFQ is bits and not packets. So there is no unfairness in the scheduling for sessions with smaller packet sizes. With WFQ, each queue is scheduled based on a computation performed on the bits of each packet at the head of the queue. Because the traffic computation is done based on stream of bits and not of packets and because what the router receives and transmits are indeed packets, WFQ implicitly increases complexity. Figure 7.8 illustrates the WFQ scheduling principle.

In Figure 7.8, all queues have the same weight, and thus the number of bytes scheduled in each scheduling turn is the same for all queues and reflects the weight value. Q0 removes three packets with a total value of X bytes, Q1 removes one packet with Y bytes, and Q2 removes two packets with Z bytes. The weight factor is effectively an allowance for how many resources can be assigned and used.

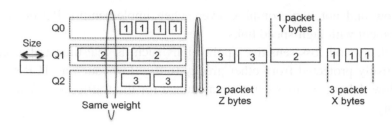

Figure 7.8   WFQ scheduling

The drawback of WFQ is that it is very resource intensive because of the bit computations. The original WFQ idea also consumes many resources because the flows are not aggregated into classes with limited queues. Instead, each flow or stream gets its own queue or buffer quota, similar to FQ. Because of these high resource demands and the complex computation needed to check the state for each flow and its packets, WFQ has been implemented more on CPU-based platforms whose queuing disciplines are based on bus-based architectures.

WFQ has the following benefit:

- The implementations based on WFQ algorithm provide service differentiation between classes and their aggregated traffic, rather than merely differentiating between individual flows. A weight allocated to each class divides the scheduling and bandwidth ratio for each class. Also, because WFQ is bits aware, it can handle packets of variable lengths.

WFQ has the following limitations:

- The original WFQ design is more of a queuing theory. The existing implementations do not follow the original concept in which each flow is allocated a weight. Instead, flows are aggregated by being classified into different service classes, and these classes are then assigned to queues.
- WFQ is extremely complicated to implement, as is FQ. Maintaining state information and computing the hash table are resource-intensive tasks.

7.8   WRR

WRR is a scheduling discipline that addresses the shortcomings of PQ and FQ. The basic concept of WRR is that it handles the scheduling for classes that require different bandwidth. WRR accomplishes this by allowing several packets

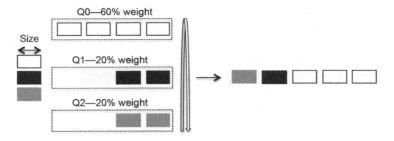

Figure 7.9   WRR scheduling

to be removed from a queue each time that queue receives a scheduling turn. WRR also addresses the issue with PQ in which one queue can starve queues that are not high-priority queues. WRR does this by allowing at least one packet to be removed from each queue containing packets in each scheduling turn. At first glance, it may seem that WRR is very similar to WFQ. The difference between the two is that WFQ services bits at each scheduling turn, whereas WRR handles packets in each scheduling turn. The number of packets to be serviced in each scheduling turn is decided by the weight of the queue. The weight is usually a percentage of the interface bandwidth, thereby reflecting the service differences between the queues and the traffic classes assigned to those queues. Figure 7.9 illustrates WRR.

As illustrated in Figure 7.9, three packets are removed from Q0 and one packet from both Q1 and Q2. The number of packets removed reflects the weight for the queue. As seen, Q0 has three times more weight than Q1 and Q2, and then it removes three times more packets each scheduling turn.

WRR has no knowledge of the true sizes of the packets in the buffers that are to be scheduled. The queues and scheduling are generally optimized for an average packet size. However, the sizes are all just estimates and have no true meaning with regard to the actual traffic mix in each queue. This operation of WRR is both an advantage and an issue. Because WRR has no complex resources that demand state computation as with WFQ, which must transform bits to bandwidth scheduling, it is fairly simple to implement WRR. The result is a solution well suited for handling a large number of flows and sessions, making WRR into something of a core QOS solution that can deal with large volumes of traffic and with congestion. The drawback of WRR is that it is unaware of bandwidth because it does not handle variable-sized packet. In terms of receiving a schedule turn, a 1500-byte packet is equivalent to a 64-byte packet. In practice, the packets' weight counts only in each round of service scheduling. When the traffic mix is

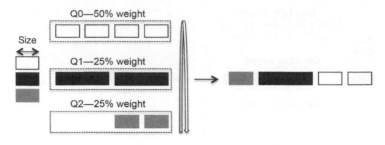

Figure 7.10 Comparing WRR with large and small packets

somewhat the same with regard to classes and queues, WRR is, over time, acceptably fair in its scheduling. However, in the short term, the differences can be great. And if traffic classes have large variations in traffic mix and traffic types and thus large differences in packet sizes, the queuing situation can become quite unfair, favoring classes that are dominated by large packets. For example, a TCP-based traffic class can gain the advantage over a real-time traffic class whose packets are relatively small, as illustrated in Figure 7.10.

In Figure 7.10, Q1 and Q2 have the same weight. However, because the packets in Q1 are twice the size of packets in Q2, in reality Q1's bandwidth rate is twice that of Q2.

WRR has the following benefits:

- The WRR algorithm is easy to implement.
- To implement differentiated service, each queue is allocated a weight that reflects the interface's bandwidth. This scheme is fair because it avoids having one queue starve another queue and because it is possible to provide service priority by allocating various weights to the queues.

WRR has the following limitations:

- Because WRR is unaware of how different packet lengths are scheduled, scheduling can be unfair when queues have different sized packets. When one queue has mostly small packets while another has mostly big packets, more bandwidth is allocated to the queue with big packets.
- Services that have a very strict demand on delay and jitter can be affected by the scheduling order of other queues. WRR offers no priority levels in its scheduling.

7.9 DWRR

With WRR for each scheduling turn, the number of packets that are granted service is based on a weight that reflects the bandwidth allocation for the queue. As discussed in the previous section, bandwidth allocation can be unfair when the average packet sizes are different between the queues and their flows. This behavior can result in service degradation for queues with smaller average packet sizes. Deficit Round Robin (DRR), or Deficit Weighted Round Robin (DWRR), is a modified weighted round-robin scheduling discipline that addresses the limitations of WRR. Deficit algorithms are able to handle packets of variable size without knowing the mean size. A maximum packet size number is subtracted from the packet length, and packets that exceed that number are held back until the next scheduling turn.

While WRR serves every nonempty queue, DWRR serves packets at the head of every nonempty queue whose deficit counter is greater than the size of the packet at the head of the queue. If the deficit counter is lower, the queue is skipped and its credit value, called a *quantum*, is increased. The increased value is used to calculate the deficit counter the next time around when the scheduler examines the queue for service. If the queue is served, the credit is decremented by the size of packet being served.

Following are the key elements and parameters in implementing DWRR that affect the scheduling service for the queues:

- Weight, which is similar to WRR algorithm weight, reflects a proportion of the bandwidth on the outgoing interface.
- The quantum translates the weight value into bytes. With each scheduling turn, a value of quantum is added in proportion to the weight. Thus, the quantum is the throttle mechanism that allows the scheduling to be aware of the bandwidth.
- The value of credits can be positive or negative. Positive credits accrue when, at the end of a scheduler turn, there are leftover bytes that were not used in the queue's scheduling turn. This value is deferred to the queue's next scheduling turn. Negative credits accrue when the queue has transmitted more than its bandwidth value in a scheduling turn, and thus the queue is in debt when the next scheduling turn comes around.
- The deficit counter, which provides bandwidth fairness, is the sum of the quantum and the credits. The scheduler removes packets until the deficit counter reaches a value of zero or until the size of the packets is larger than the remaining deficits. But if the queue does not have enough deficits to

schedule a packet, the value of the deficit is retained until the next scheduling round, and the scheduler moves to the next queue.

Let us go through DWRR in practice to illustrate its attributes. Consider a setup as shown in Figure 7.11 with three queues with different weights that reflect the proportion of bandwidth of outgoing interface. Q0 has 50% of the bandwidth, and Q1 and Q2 each have 25% of the bandwidth. To simplify the example, all queues have an equal length. The queues have variable-sized packets and different numbers of packets. At each scheduling turn, credits are added to the quantum number of each queue that reflects the weight as a quota of the bandwidth.

Figure 7.12 shows the first scheduling turn and the process of dequeue for Q0. In Figure 7.12, we see that Q0 has three packets, two that are 300 bytes and one that is 500 bytes. Each scheduling turn, a quantum value of 1000 is added to Q0. In this turn, then, the deficit counter is 1000, which means that 1000 bytes can be subtracted from the queue. The 300-byte and 500-byte packets can be transmitted because they consume 800 credits from the 1000-credit bucket.

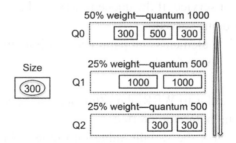

Figure 7.11   DWRR scheduling configuration

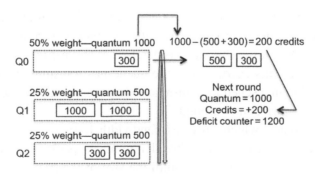

Figure 7.12   DWRR scheduling, turn 1, Q0

But the last 300-byte packet cannot be transmitted in this turn, because 300 are bigger than the remaining 200 credits $(1000-800=200)$. The result is that the scheduling jumps, in order, to the next queue with packets inside and the 200 credits for Q0 are saved for next scheduling turn. Now let's look at the next queue in order, Q1, as illustrated in Figure 7.13.

Q1 contains two 1000-byte packets, but the quantum per scheduling turn is 500 credits. This means that no packets can be transmitted in this scheduling cycle because not enough credits exist $(500<1000)$. The result is that the deficit counter for Q1 is incremented with 500 credits for the next scheduling round. This example illustrates the power of deficit-style algorithms compared with plain round-robin systems. With DWRR, Q1 is actually punished because the available quantum and credit values are smaller than the actual packet sizes. This situation is proof that DWRR scheduling is aware of variable packet sizes, not just the number of packets to be queued. Let us move to the next queue in order, Q2, shown in Figure 7.14.

Q2 has a quantum value of 500, and because the packet at the head of the queue is 300 bytes, it is transmitted. The value of 300 is subtracted from the 500, resulting in 200 credits that are saved for the next scheduling turn. The scheduling wheel has now completed one full turn, and we are back at Q0, as illustrated in Figure 7.15.

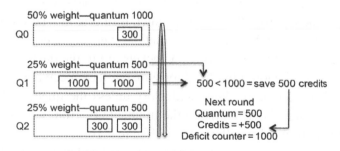

Figure 7.13   DWRR scheduling, turn 1, Q1

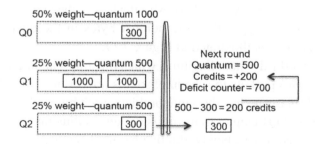

Figure 7.14   DWRR scheduling, turn 1, Q2

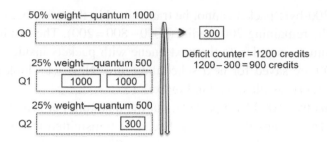

Figure 7.15 DWRR scheduling, turn 2, Q0

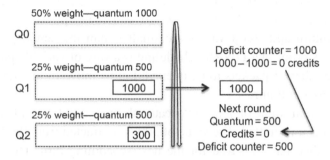

Figure 7.16 DWRR scheduling, turn 2, Q1

The Q0 deficit counter is now 1200. The 300-byte packet is transmitted and its value is subtracted from the deficit counter, resulting in a new value of 900 credits (see Figure 7.15). While implementations can vary, a common behavior is to set the deficit counter to zero if no packets are in the queue. Queues that behave this way are honored by the algorithm, but they receive no extra points for no work. You cannot get fame and glory unless the queue is filled with packets. Now the scheduler returns to the troubled Q1, from which no packets were transmitted in turn 1, as illustrated in Figure 7.16.

Finally, there is some good news for Q1. The updated quantum value is 500, and with the 500 credits saved from cycle 1, for a total of 1000 credits, Q1 can transmit one of the 1000-byte packets. Moving to Q2, the news is all good. The deficit counter is far bigger than the number bytes in the queue, so the last packet is transmitted, as illustrated in Figure 7.17.

Let us look at a third scheduling turn, which illustrates an interesting DWRR scenario. Q0 has no packets to schedule, so it is not serviced and the scheduler moves to Q1. This queue has the same problem, with big packets eating up the deficit counter, and Q1 cannot transmit its packet, as illustrated in Figure 7.18.

Scheduling turns, that is, bandwidth that has not been used by one queue can be shared with other queues. So for poor Q1, this is good news. The extra

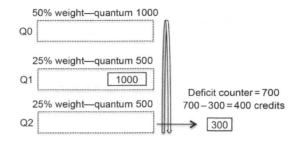

Figure 7.17 DWRR scheduling, turn 2, Q2

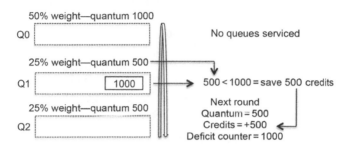

Figure 7.18 DWRR scheduling, turn 3, Q1

available quantum of credits from Q0 is shared with other queues that have packets to be transmitted, with the leftover quantum divided between the queues based on their weight: the more weight, the more of the extra quantum credits are allocated. Because Q2 had no packets in its buffer, the entire leftover quantum for Q0 is given to Q1. This is illustrated in Figure 7.19.

How does a deficit-based system drop packets? The previous example did not explicitly examine this. In the packet-dropping schemes described earlier in this chapter, packets can be dropped at the head of the queue when packets remain in the queue too long because of congestion, and packets can also be dropped from the tail when they cannot be placed into a full queue. This is also the process for DWRR. In the previous example, for ease of understanding, the credits were set to a low value. For the most part, the credit value reflects the MTU size of the outgoing interface. Thus, the credit refill rate should keep up with the queue refill rate to avoid a complete forwarding stalemate in the queue.

An alternative way to describe the state in which no scheduling occurs is with the concept of negative versus positive credit states. When a queue enters a negative state, it does not receive any service because it exceeded its rate credit in its last scheduling turn. A queue can enter negative credit territory if the last packet could not be transmitted because of insufficient credits, resulting in a deficit counter that is negative. Once the queue is in this negative credit state, it

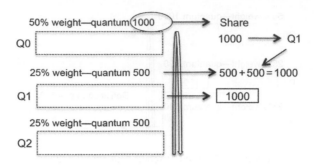

Figure 7.19   DWRR scheduling, turn 3, leftover bandwidth

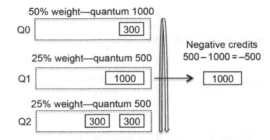

Figure 7.20   DWRR scheduling, turn 1, Q2 with negative credits

cannot schedule packets until the quantum of credits added returns the currently negative deficit counter value to a positive number. The only way to get into positive credit territory is to have the queue build up enough credits to be able to schedule packets again. In summary, a queue does not receive any scheduling slots unless the deficit counter is greater than zero.

Let us consider the earlier scenario with Q1. Here, Q1 is implicitly punished because it has attempted to use more than its quantum of credit. The result is that in the scheduling turn 1, no packets are transmitted because the quantum values are less than the number of bytes in the queue. In such a scenario, packets drops occur at the head of the queue because the packets stay in the queue too long and age out. In the case of severe congestion, clearing of the queue cannot keep up with the rate at which the queue is filled.

With the logic of negative credits, a queue is allowed to have a negative value for the deficit counter. Figure 7.20 shows that Q1 receives a negative credit value during the first scheduling turn because the 1000-byte packet is bigger than the quantum of 500. The result is a deficit counter value of −500.

At the next scheduling turn for Q1, the state is negative, and the quantum of 500 is not higher than the deficit counter, so no packets are scheduled. Because the deficit counter is less than zero, the scheduling algorithm then skips to next queue. The good news, however, for Q1 is that at the third scheduling turn, the queue has a positive credit state and could schedule packets again. Whether Q1 actually drops the 1000-byte packet while it is waiting to receive a scheduling turn depends on the buffer depth and the fill rate of the queue. The negative versus positive credit state is a way to honor well-behaved queues and punish poorly behaved ones.

DWRR has the following benefit:

• The DWRR algorithms limit the shortcomings with WRR because they are a more modern form of WFQ that incorporates scheduling aware of bits and packets.

DWRR has the following limitation:

• Services that have very strict demand on delay and jitter can be affected by other queues by the scheduling order. DWRR has no way to prioritize its scheduling.

7.10 PB-DWRR

Introducing the deficit counter allows the WRR algorithm to be aware of bandwidth and improves the fairness. But for certain traffic types, fairness is not the desired behavior. What is needed is a priority scheduling similar to PQ, but that preserves the benefits of DWRR. To achieve predictable service for sensitive, real-time traffic, a priority level for scheduling needs to be introduced. By enabling strict priority or by offering several priority levels and using DWRR scheduling between queues with the same priority levels, service assurance with regard to delay and loss protection can be achieved for demanding traffic types, such as voice and real-time broadcasting. To illustrate this, consider three queues, one of which has a priority of strict high, as shown in Figure 7.21.

When service cycles are scheduled, Q2, which has strict-high priority, always has preference over the two low-priority queues. In each scheduling turn, Q2 is drained first. If the strict-high queue is filled faster than it is cleared, the result is that the packets in Q0 and Q1 become stale and can age out because these queues are not behaving within their defined weights. This aging out of packets is a side effect of the fact that a strict-high queue never goes into negative credits. However, a strict-high priority queue system can work if the rate of packets entering Q2 is controlled, limited, for example, to 20% of the total bandwidth. Then the credit state is reflected accordingly. One way to control the rate into a queue is by adding a policer that limits the rate by dropping packets that exceed

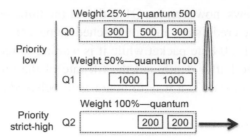

Figure 7.21   PB-DWRR scheduling, with one strict-high priority queue

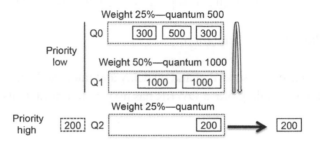

Figure 7.22   PB-DWRR scheduling, policed priority-high queue

a defined rate limit. This scheme avoids having the strict-priority queue enter a runaway state in which it monopolizes all the scheduling cycles. Another way to control the rate into the queue is to control the delay for the queue. Figure 7.22 illustrates rate control. Assigning a weight of 25% of the scheduling bandwidth to Q2 prevents more than one packet from being in Q2 at a time; as a result, Q2 drops one of the 200-byte packets at the tail of the queue.

Instead of using a strict-high priority queue, an alternative way to rate limit the queue is to use an alternating priority scheme. This method allows the priority-high queue to be scheduled between every other queue. For example:

Queue 0, Queue 2, Queue 1, Queue 2, Queue 0, Queue 2, …

An alternating priority mode can, however, cause some unpredicted delay because the other queues could introduce delay depending on the traffic that they schedule. Also, with alternating priority, the rate of the queue is not controlled. Another way to handle fairness is to have queues behave according to their weight, without caring about priority if they exceed their weight. This solution prevents the previously illustrated scenario of runaway higher-priority queues causing packets in lower-priority queues to become stale. Consider the PB-DWRR configuration and queue status in Figure 7.23.

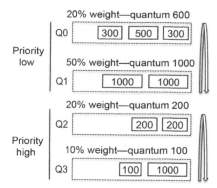

Figure 7.23   PB-DWRR scheduling with two priority-high queues

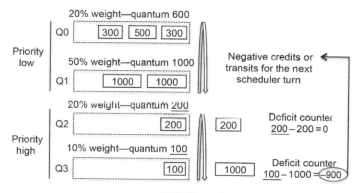

Figure 7.24   PB-DWRR scheduling, turn 1, Q2 and Q3

   This scenario shows four queues, with two different priority levels. The different priority levels means that there are multiple levels of scheduling. Q2 and Q3 have a high-priority level. If there are packets in these two queues and if they have enough credit state, they are serviced before the two low-priority queues.

   Let us evaluate the first scheduling turn. Because DWRR separates queues with the same priority level, Q2 is scheduled to receive service in the first cycle. The quantum for Q2 is 200. Thus, its credit state is 200 and it can clear one 200-byte packet from the queue. Q2's deficit counter value is now 0, and as explained earlier, the deficit counter must be greater than zero to receive service in the same scheduling cycle. The next queue visited is the next priority-high queue, Q3. This queue schedules one packet but then enters into a negative credit state because of its quantum. Q3's credit state is 100. When removing 1000 bytes from the queue, the value of the deficit counter becomes –900 (see Figure 7.24).

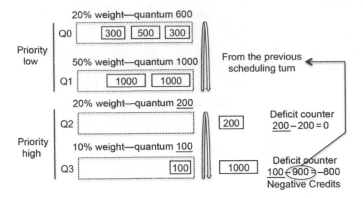

Figure 7.25   PB-DWRR scheduling, turn 2, Q2 and Q3

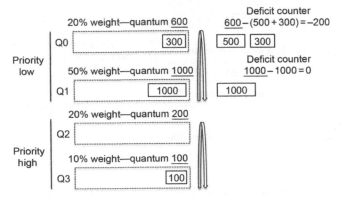

Figure 7.26   PB-DWRR scheduling, turn 2, Q0 and Q1

At the next scheduling turn, Q2 is visited again because its priority level is high and because it contains packets. The quantum is 200 bytes, which is also the number of bytes waiting to be cleared from the queue, as illustrated in Figure 7.25.

Now something interesting happens: Q3 is in negative credit state. This means it gets no scheduling cycles because the deficit counter is not greater than zero. The result is that turn 2 in the scheduling now jumps to the low-priority queues. Because Q1 has a higher weight than Q0, the scheduling cycle services Q1, which removes one 1000-byte packet from the queue, leaving its deficit counter at a value of 0. The next queue to be scheduled in turn 2 is Q0, and it removes 800 bytes before running into a negative credit state, as illustrated in Figure 7.26.

Now we have a scenario in which one of the priority-high queues, Q3, and the low-priority Q0 are in negative credit state. If we honor the state of the queues, in the next scheduling turn, the queue to get scheduling service is Q1 because it is not in negative credit state. Because Q2 has no packets in the queue, the two queues in negative credit state can use Q2's weight and thereby update their credit states based on available weights. Because a queue can receive service only once the deficit counter is greater than zero, the queue needs to be bypassed in some scheduling rounds to increase their credits. If no other queues with positive credit state have packets in their queues, this refill of credits happens faster because the punished queues can reuse the available weights not being used by other queues.

What is the most efficient way to schedule queues with the same priority level? A good design is probably a combination of strict high and DWRR scheduling between queues with the same level of priority. This is a good design in the core of the network for situations in which the strict-priority queue may need more bandwidth for a short time because of a network failure and the resultant reconvergence. Limiting real-time traffic should instead be done on the edge of the network, and using a combination of strict high and DWRR schedule is likely also a good design. But if there are multiple priority-high queues, some restrictions or control, for example, DWRR, are needed to avoid possible resource clashes between multiple high-priority queues.

PB-DWRR has the following benefit:

- PB-DWRR incorporates the lessons learned from most other queuing algorithms. It incorporates most of the features introduced by these algorithms, such as byte deficit scheduling and priority levels.

PB-DWRR has the following limitation:

- It is not a standard, so all routers and switches that implement it may not act in a similar fashion. PB-DWRR is highly dependent on the hardware it runs on, the available resources, and of course, on the actual vendor implementation.

7.11 Conclusions about the Best Queuing Discipline

As of this writing, the most popular queuing and scheduling mechanism is PB-DWRR. The reason behind its success and acceptance in the industry is that PB-DWRR is built on the lessons learned from its predecessors. This chapter has described PB-DWRR concepts. However, actual vendor implementations may vary slightly. As with any queuing and scheduling mechanism, there is always a dependency on resources and hardware. However, one piece of the queuing and scheduling puzzle is still missing: the WRED profiles. We discuss these in the next chapter.

Further Reading

Clarke, P., Saka, F., Li, Y.-T. and Di Donato, A. (2004) *Using QoS for High Throughput TCP Transport over Fat Long Pipes*. London: University College London.

Croll, A. and Packman, E. (2000) *Managing Bandwidth: Deploy QOS in Enterprise Networks*. Upper Saddle River, NJ: Prentice Hall.

Goralski, W. (2010) JUNOS Internet Software Configuration Guide: "Class of Service Configuration Guide," Release 10.1, Published 2010. www.juniper.net (accessed September 8, 2015).

Heinanen, J., Baker, F., Weiss, W. and Wroclawski, J. (1999) RFC 2597, Assured Forwarding PHB Group, June 1999. https://tools.ietf.org/html/rfc2597 (accessed August 18, 2015).

Jacobson, V., Nichols, K. and Poduri, K. (1999) RFC 2598, An Expedited Forwarding PHB, June 1999. https://tools.ietf.org/html/rfc2598 (accessed August 18, 2015).

Nichols, K., Blake, S., Baker, F. and Black, D. (1998) RFC 2474, Definition of the Differentiated Services Field, December 1998. https://tools.ietf.org/html/rfc2474 (accessed August 18, 2015).

Postel, J. (1981) RFC 793, Transmission Control Protocol—DARPA Internet Protocol Specification, September 1981. https://tools.ietf.org/rfc/rfc793.txt (accessed August 18, 2015).

Semeria, C. (2000) Internet Backbone Routers and Evolving Internet Design, White Paper, Juniper Networks. www.juniper.net (accessed August 18, 2015).

Semeria, C. (2001) Supporting Differentiated Service Classes: Queue Scheduling Disciplines, White Paper, Juniper Networks. www.juniper.net (accessed August 18, 2015).

8

Advanced Queuing Topics

Chapter 7 focused on queuing and scheduling concepts and how they operate. This chapter discusses more advanced scenarios involving overprovisioning and guaranteed rates that are used in large-scale queues to differentiate shaping rates. This chapter touches on the debate about the optimal size of the queues and different memory allocation techniques. Finally, the chapter details the famous (or infamous) Random Early Discard (RED) concept.

8.1 Hierarchical Scheduling

The techniques in Chapter 7 described single-level port scheduling. But modern QOS scenarios and implementations need to support multiple levels of scheduling and shaping. A situation when this is common is a subscriber scenario that provisions end users. Consider the topology in Figure 8.1.

The Broadband Service Router (BSR) in Figure 8.1 has a 10-Gbps Gigabit Ethernet link to the switch. The two connections between this switch and the two DSLAM switches are both Gigabit Ethernet links. The subscribers have traffic contracts with different services and rates. To be able to have a functional traffic provisioning model, rates to and from the subscribers need to accommodate service and contracts. Traffic from the subscribers must be shaped to align with the service contract, and traffic to the subscribers must be shaped according to the supported model. To meet these conditions, a hierarchical shaping model must be created on the BSR router link to and from the aggregation switch. At the port level, traffic to the switch is first shaped to 1 Gbps on the two SVLANs,

QOS-Enabled Networks: Tools and Foundations, Second Edition. Miguel Barreiros and Peter Lundqvist.
© 2016 John Wiley & Sons, Ltd. Published 2016 by John Wiley & Sons, Ltd.

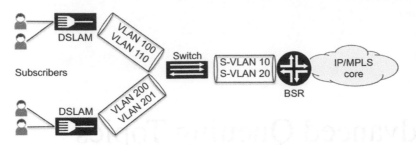

Figure 8.1   Aggregated hierarchical rate provisioning model

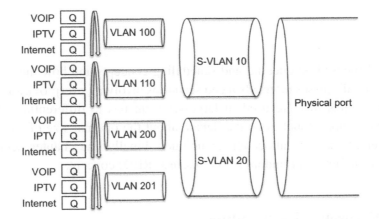

Figure 8.2   Hierarchical scheduling model

SVLAN 10 and SVLAN 20 in Figure 8.1. The next layer of shaping occurs on the VLANs that exist within each SVLAN. For example, some users might have 12 Mbps contracts, while others have 24 Mbps. In Figure 8.1, these VLANs are 100 and 110 on the upper DSLAM switch and 200 and 201 on the lower DSLAM.

The next level of scheduling is the ability to provide a differentiated model to classify different types of traffic, for example, VOIP, IPTV, and best-effort (BE) Internet traffic, by scheduling each into a different queue. Figure 8.2 shows the hierarchical scheduling and shaping model for this scenario.

Figure 8.2 shows two levels of aggregated shaping and scheduling. Routers that support hierarchical scheduling and shaping face issues not only with back pressure in levels from the port that carries SVLAN and VLAN toward the subscribers, but also, by the nature of the service, they need to support a large amount of traffic and they need to be able to scale.

QOS scheduling on a large scale, with a large number of units and schedulers, has to be able to propagate important traffic. Consider a case in which you have

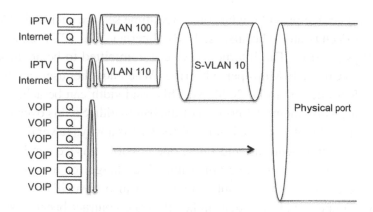

Figure 8.3 Propagation of priority queues

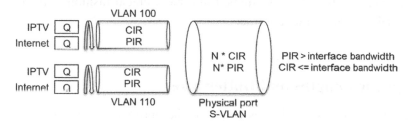

Figure 8.4 The PIR/CIR model

1000 subscribers on an interface, with one logical interface (i.e., one unit) per subscriber. In the case of congestion, the scheduling algorithm establishes that each unit is visited in turn in case they have packets in their queues. This mechanism can result in variable delay, depending on how many users have packets in their queues. Variable delay is not desirable for users who have triple-play services with, for example, VOIP. The scheduling must be able to propagate information about the priority of the queues before all the subscriber queues are visited in order. Figure 8.3 shows how all user queues with high-priority VOIP traffic are scheduled before each individual user is visited.

A second scenario for aggregated hierarchical rate and scheduling models involves overprovisioning. A common scenario is to balance oversubscription against guaranteed bandwidth for the users. This concept is most often referred to as Peak Information Rate (PIR) versus Committed Information Rate (CIR). The sum of all the CIRs on an interface cannot exceed the interface rate or the shaped rate, but the PIR is allowed to exceed these rates. If all the bandwidth is not used, subscribers and shaped VLANs are allowed to reuse that bandwidth, as illustrated by Figure 8.4.

Consider a service provider who sells a real-time service such as IPTV Video on Demand (VOD) and who also sells BE Internet. The IPTV bandwidth is about 6 Mbps per user. The Internet service is committed to be about 4 Mbps. The result is a CIR of 10 Mbps per subscriber. When no other users are transmitting, the PIR is allowed to be 20 Mbps. This bandwidth can be achieved if no other users are using their CIR portion of the bandwidth. So this overprovisioning model has a PIR of 2:1. Traditionally, the CIR is always lower than the PIR. Using hierarchical scheduling and rates, one possible service can be to shape Internet service for users to a certain value. The charge for Internet access is a fixed price each month. For subscribers who also have IPTV service, the allowed VOD rate could be much higher than the contract because sending a movie over the VOD service results in direct payment to the operator. Therefore, the VOD rate is scheduled and shaped in a hierarchical fashion, separate from the subscriber Internet rate contract.

8.2   Queue Lengths and Buffer Size

One debate that comes and goes concerns the optimal size of the queue. It is sometimes called the small versus big size buffer debate. This debate can be viewed from a macro level or from a micro level.

First, let us consider the macro debate, which centers on the question of how to design congestion avoidance in the big picture. On one side of this debate are the supporters of big buffers. The designers who favor this approach argue that in packet-based networks on which QOS is implemented using per-hop behavior (PHB), routers and switches make all the decisions on their own, with no awareness of the end-to-end situation. Hence, here an end-to-end QOS policy is simply not doable. Traffic destinations change as the network grows or shrinks on daily basis. Thus, the rerouting that occurs as the network continually changes size invalidates any attempt to predesign paths, with the result that any traffic modeling easily breaks down. In this design, end-to-end RTT estimates are hard to predict, so it is a good idea for all resources to be available. The solution is to go ahead and add memory to expand the capability of storing traffic inside a router. One benefit of this is that big buffers can temporarily absorb traffic when traffic patterns or traffic paths change. The idea is that a combination of well-designed QOS policy and tuned bandwidth and queue length works only if the traffic patterns and paths are somewhat static and easy to predict. In the macro situation, rather than having an end-to-end QOS policy, the QOS policy can be adapted to define specific traffic classes and to align with the capabilities of the

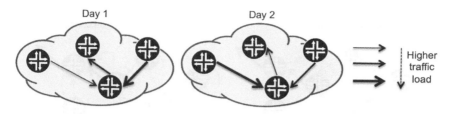

Figure 8.5 Every router stands on its own regarding QOS

routers. If one router can support a queue length of hundreds of milliseconds, while another router in the traffic path can support up to seconds of traffic buffering, the QOS policy can be tailored to each router's capabilities. In this way, the policies can be based on each router's functions and capabilities, because traffic and other conditions can vary widely from router to router and day by day, making end-to-end policies ineffective or impossible. Figure 8.5 illustrates the main argument for defining QOS policies on each individual router.

The opposing view is that end-to-end QOS policy is the only way for QOS to be effective, and hence buffers must be tuned to be a part of an estimated end-to-end delay budget. This means that for the possible congestion paths in the network, prioritizing the expected traffic is not enough. It is also necessary to tune the buffers to be able to carry the traffic and at the same time be within the end-to-end delay figures. The estimation of what a fair share of the path and bandwidth is needs to take into account both the steady-state situation and predicted failure scenarios along the traffic paths and must also anticipate possible congestion points. The rule to be followed with this approach is to control the rate of traffic entering the network. The amount of traffic that needs to be protected cannot be greater than what the network can carry. To achieve this service, the convergence and backup paths need to be calculated. A common way to control the paths is to use RSVP LSP signaling, implementing object constraints regarding bandwidth and ERO objects to control the end-to-end delay and bandwidth, because LSPs allow complete control over the path that the traffic takes (see Figure 8.6).

When considering these two approaches, the reality is probably that good queue length tuning needs to use a little of both alternatives. Real-time applications obviously need to have reliable control of traffic paths, which means that the number of congestion points need to be minimal and the end-to-end delay, in milliseconds, needs to be predictable. For these applications, designers generally think that buffers should be small and paths should be well designed. Also, in the case of a network path outage, equivalent convergence backup paths should be included in the design.

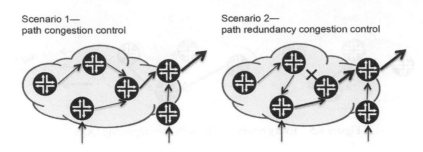

Figure 8.6   End-to-end QOS policy for certain traffic classes

Here is a concrete example of balancing the queue length and buffer size. A transit router has four 10-Gigabit Ethernet (10GE) links. (A 10GE link is 10000 Mbps.) Transforming this to bytes gives 1000/8 or 1250 Mbps. Transforming this to milliseconds gives 1250×1000 or 1250 kB. This means that 1 ms of traffic is 1250 kB. If the router supports 100 ms of traffic buffering, the buffer allocated to the 10GE interface is 100×1250 kB or 125 000 000 bytes of memory.

Under steady conditions, the estimate is that a router has one duplex path for transit voice traffic and the volume for one direction is 500 Mbps. If a reroute event occurs, one more duplex path of voice traffic transits the router. In the case of congestion, the core link may suddenly find itself carrying 1000 Mbps of voice traffic or 125 kB/ms. This traffic volume can be handled well by a router capable of buffering 100 ms of traffic on the 10GE interface.

In the first design approach, in which each router handles QOS on its own, the calculation for QOS ends here. The second design approach demands that buffers be adapted to the end-to-end RTT and that backup paths be well controlled, for example, by link-protection schemes. If a reroute scenario demands bandwidth and ERO cannot be achieved, the routing and forwarding tables cannot deliver the traffic. The result is that RSVP would resignal the ingress and egress points of the LSP to achieve the ERO requirements.

One possible design compromise is to use RSVP LSP paths for loss-sensitive traffic to control the buffer length and paths, effectively establishing a predictable delay and handling the bulk of the BE traffic with the available buffer space on each router in the network.

Now, let us consider the micro view, which considers the best behavior for the majority of applications and protocols regarding buffer tuning. Most designers agree that delay-sensitive and loss-sensitive traffic favors a design with smaller buffer lengths. The big difference between the micro and macro views is probably the dimensioning of the queue size for the TCP BE traffic. That is, should we use short queues so that bad news can be relayed quickly? The result would be earlier retransmissions or even the avoidance of too much congestion from transmitting hosts. In theory, retransmissions within a session are limited to a few packets because the queue should not be very full. The whole purpose of using smaller queue sizes is to avoid a "bigger mess" caused by delays in delivering bad news. With small queues, the delay is relatively low and jitter is not much of an issue, but the volume of packet retransmissions is higher. Web applications that are short lived are more sensitive to delay because response time must be quicker since there is not much CWND buildup and not many bytes of traffic are transferred. On the other hand, should the queue be big to be able to smooth out bursts, but at the cost of increased delay or even a huge jitter? Having large queues would avoid retransmissions and allow smoother traffic delivery, and they ignore the delay factor because no delay occurs if the sessions exist just to transfer a file.

You can never have enough buffers to smooth the bursty traffic of all possible BE users in today's Internet or large enterprise networks. Consider TCP-based traffic that is elastic with regard to its congestion and receiving windows. The memory necessary to be able to buffer all user streams is equal to all possible active TCP sessions, and the sessions of traffic must be multiplied by the maximum size of each session's TCP window. In most situations, providing this amount of memory is unrealistic.

8.3   Dynamically Sized versus Fixed-Size Queue Buffers

As we have discussed earlier, other queues can be configured to use bandwidth that is not being used by higher- priority or active queues. This means, for example, that when saturation and packet drops are expected with a BE queue, the queue can receive better service if no assured-forwarding or expedited-forwarding packets need the allocated bandwidth and scheduling time. BE service is normally not rate-limit or policed to conform to a specific rate, so if a BE queue gets more scheduling time because other queues have nothing to remove, this is simply the nature of BE traffic and oversubscription.

But the question arises about whether you should share buffers also. That is, should you allow the queue size memory quota to be reused if it is not needed by other queues? The advantages with this, of course, are being able to absorb a greater burst of packets and to reuse the memory space. Designers who favor this approach are supporters of having routers with large queuing buffers establishing their own QOS policy.

The term most commonly used to describe this method is memory allocation dynamic (MAD). While the logic of implementing MAD seems fairly

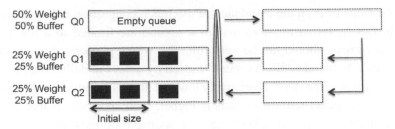

Figure 8.7 Dynamic buffer allocation example

simple, it is actually relatively complex because the queue length status must be taken into consideration when scheduling turns allocate bandwidth based on the deficit counter and credit state. An additional algorithm is required to reallocate buffers for the queues currently not transmitting to add entries to those queues.

Figure 8.7 illustrates how MAD operates. Queue zero (Q0) contains no entries, while Q1 and Q2 are approaching a trail-drop situation because these buffers are close to overflowing. The extra scheduling (bandwidth) cycles and buffer space help the queues when they are highly saturated.

It is worth noting that some designers do not like being able to share both buffer space and bandwidth, because the delay obviously cannot be predicted. The most common design is to allocate a queue for a BE traffic type class, thereby offering the ability to get more buffering space in case of need, but to have a more traditional static buffer length for expedited-forwarding traffic class as a way to keep the delay values below a strict margin.

8.4 RED

Throughout this book, we have highlighted the applicability of RED and weighted RED (WRED) for TCP flows and as a way to differentiate between different traffic types inside a queue. Let us now dive into the concepts behind RED and its variations. Queuing and scheduling handle bandwidth and buffering once congestion exists. Once packets are in the queue, they are either transmitted after spending some time in the queue or are aged out and dropped because the queue has no scheduling cycles. But what about dropping a packet before it enters a queue or dropping a packet in the queue before it gets aged out? What congestion avoidance tools are available? There are two main traffic congestion avoidance tools, policers, which have been described earlier in Chapter 6, and RED.

As described earlier, policing can be "hard" (i.e., packets are dropped if traffic exceeds a certain rate) or "soft" (packets are remarked if traffic exceeds a certain rate). Hard policing can ensure that delay-sensitive traffic entering at the edge of the network does not exceed the rate supported across the core. It can also be used to rate-limit BE traffic that is being transmitted into the network maliciously, for example, during an ICMP attack. And policing can be used with RED. So what is this RED?

RED is a tool that provides the ability to discard packets before a queue becomes full. That is, RED can drop packets in the queue before the congestion becomes excessive, before packets are aged because the queue is overflowing, and hence before traffic stalls. Floyd and Jacobson [1] first proposed RED in 1993 in their paper *Random Early Detection Gateways for Congestion Avoidance*. The idea behind RED is to provide feedback as quickly as possible to responsive flows and sessions, which in reality means TCP-based sessions. The thinking behind RED is to flirt with a TCP congestion avoidance mechanism that kicks in when duplicate ACKs are received, acting by decreasing the congestion window (CWND). The packet drops are not session related; rather, they are distributed more fairly across all sessions. Of course, when a session has more packets in the buffer, it has a greater chance of being exposed to drops than lower-volume sessions.

Let's first examine the default queue behavior, tail drop. A queue has a specific length, which can be translated into how many packets it can accommodate. When a packet arrives to be put into the queue, it is either dropped if the queue is full or stored in the queue. This behavior is somewhat analogous to a prefix-limiting policy that can be applied to any routing protocol, in which new routes are accepted until a limit is reached, and then no new routes are accepted.

In Figure 8.8, the vertical axis represents a drop probability, and the horizontal axis represents the queue fill level, and both parameters can range from 0 to 100%. As illustrated here, the drop probability is always 0% unless the queue is full, in which case the probability jumps straight to 100%, meaning that once the queue is full, all packets are dropped. This standard tail-drop behavior makes sense: why drop a packet if the queue has room to store it?

RED allows the standard tail-drop behavior to be changed so that a packet can be dropped even if the queue is not full (see Figure 8.9). It sounds somewhat bizarre that the queue can store the packet but that the incoming packet can still be dropped.

Two scenarios can justify this behavior. The first, present with TCP sessions, is a phenomenon called the TCP synchronization issue, in which massive traffic loss triggers the TCP back-off mechanism. As a result, sessions enter the initial

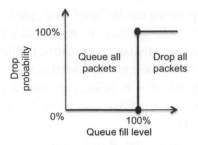

Figure 8.8   Tail-drop behavior

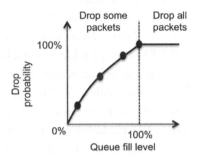

Figure 8.9   RED drops

state of TCP slow start, and they all start to ramp up their congestion windows simultaneously. The second scenario occurs when different types of traffic share the same queue and differentiating between the traffic types is necessary.

8.5   Using RED with TCP Sessions

As discussed earlier, TCP protocols have the concept of rate optimization windows, such as congestion windows and receiving windows, which can be translated into the volume of traffic that can be transmitted or received before the sender has to wait to receive an acknowledgment. Once an acknowledgment is received, data equal to the window size accepted on the receiver's and the sender's congestion windows, minus the still unacknowledged data, can be transmitted. The details of the TCP session's behavior are detailed in Chapter 4.

But let us recap briefly: the source and destination always try to have the biggest window size possible to improve the efficiency of the communication. The session starts with a conservative window size and progressively tries to increase it. However, if traffic is lost, the size of the sender's and receiver's congestion windows is reduced, and traffic whose receipt was not acknowledged is

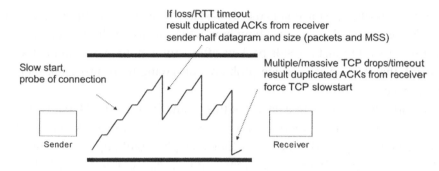

Figure 8.10 TCP congestion mechanism

retransmitted. The window size reduction is controlled by the TCP back-off mechanism, and the reduction is proportional to the amount of the loss, with a bigger loss leading to a more aggressive reduction in the window size. A severe traffic loss with timeouts causes the window size to revert to its original value and triggers a TCP slow-start stage (Figure 8.10).

The question is, how can RED usage improve this behavior? First, let us identify the problem clearly: massive traffic loss leads to a drastic reduction in window size. When is traffic loss massive? When the queue is full.

All active TCP sessions simultaneously implement their TCP back-off algorithms, lowering their congestion window sizes and reducing the volume of offered traffic, thus mitigating the congestion. Because all the sessions then start to increase their congestion window sizes roughly in unison, a congestion point is again reached, and all the sessions again implement the back-off algorithm, thus entering a loop situation between both states. This symptom is most often referred as TCP synchronization.

In this situation, RED can be used to start discarding traffic in a progressive manner before the queue becomes full. Assume that as the queue reaches a fill level close to 100%, a drop probability of X% is defined. This translates into "The queue is close to getting full, so out of 100 packets, drop X." Thus, before reaching the stage in which the queue is full and massive amounts of traffic start to be dropped, we start to discard some packets. This process provides time to allow the TCP sessions to adjust their congestion window sizes instead of reverting to their initial congestion window sizes to the slow-start stage value.

That's the theory. But does RED work? There has been a debate about how useful RED has been in modern networks. The original idea design by Floyd and Jacobson [1] focused on long-lived TCP sessions. For example, a typical file transfer, such as an FTP session, reacts by lowering the congestion window size when it receives three or more duplicate ACKs. Users are rarely aware of the event unless the session enters the slow-start phase or stalls. But the explosion of HTTP has surfaced questions about how RED can affect short-lived sessions. A web session often just resends packets after a timeout to avoid having the end user think that there were service problems or having them just click more on the web page.

It is worth discussing some basics at this point. First, for RED to be effective and keep the congestion window within acceptable limits, sessions must be long lasting. That is, they need to be active long enough to build up a true congestion window. By simply dropping sessions with limited congestion windows in response to HTTP continuation packets, RED probably cannot be effective with regard to tweaking the TCP congestion mechanism. Sessions must have a fair number of packets in the buffer. The more packets a specific session has in the buffer when the congestion is large, the greater the chance that the session experiences drops. Second, modern TCP stacks are actually very tolerant to drops. Features such as selective acknowledgment (SACK), fast retransmit, and fast recovery allow a session to recover very quickly from a few drops by triggering retransmission and rapidly building up the congestion window again.

The trend now is toward long-lasting sessions, which are used commonly for downloading streamed media, "webcasting," and point-to-point (P2P) file sharing. At the same time, current TCP stacks can take some beating for a short period without significantly slowing down other sessions. To ramp down sessions, either many sessions must be competing, thus preventing any buildup of congestion windows, or the RED profiles must be very aggressive and kick in relatively early before congestion builds up in the queue. If the "sessions" were UDP based, RED would not be effective at all except possibly to shape the traffic volume, which is likely outside the original RED design. If RED profiles just provide alternate profiles for handling how long such packets can stay in the buffer, RED can be used for UDP traffic by applying several tail-drop lengths. But again, this use does not follow the original definition of RED.

8.6   Differentiating Traffic inside a Queue with WRED

When different types of traffic share the same queue, it is sometimes necessary to differentiate between them, as discussed in Chapters 2 and 3. Here, one method used frequently is WRED. WRED is no different than applying different RED profiles for each code point (see Figure 8.11).

To provide a concrete scenario, suppose in-contract and out-of-contract traffic are queued together. While queuing out-of-contract traffic can be acceptable

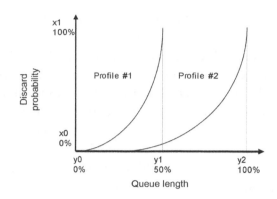

Figure 8.11   Multiple RED drop levels in the same queue

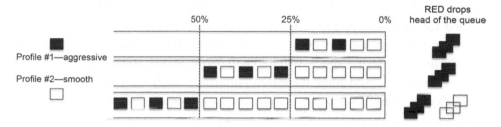

Figure 8.12   Aggressive RED drop levels for out-of-contract traffic

(depending on the service agreement), what is not acceptable is for out-of-contract traffic to have access to queuing resources at the expense of dropping in-contract traffic. Avoiding improper resource allocation can be done by applying an aggressive RED profile to out-of-contract traffic, which ensures that while the queue fill level is low, both in-contract and out-of-contract traffic have access to the same resources. However, when the queue fill level increases beyond a certain point, queuing resources are reserved for in-contract traffic and out-of-contract traffic is dropped. Figure 8.12 illustrates this design for an architecture that drops packets from the head of the queue.

In effect, this design creates two queues in the same physical queue. In this example, packets that match profile 1, which are marked with a dark color, are likely to be dropped by RED when they enter the queue, and they are all dropped when queue buffer utilization exceeds 50%. Packets matching profile 2, which are shown as white, have a roller-coaster ride until the queue is 50% full, at which point RED starts to drop white packets as well.

The classification of out-of-contract packets can be based on a multifield (MF) classifier or can be set as part of a behavior aggregate (BA). For example, the service of a BE service queue is lower than that for best-effort (LBE).

If ingress traffic exceeds a certain rate, the classification changes the code point from BE to LBE, effectively applying a more aggressive RED profile. Typically, in-contract and out-of-contract traffic share the same queue, because using a different queue for each may cause packet-reordering issues at the destination. Even if RED design mostly focuses on TCP, which has mechanism to handle possible packet reordering, having the two types of traffic use different queues can cause unwanted volumes of retransmissions and unwanted amounts of delay, jitter, and buffering on the destination node.

An alternative design for having multiple traffic profiles in the same queue is to move traffic to another queue when the rate exceeds a certain traffic volume. The most common scenario for this design is the one discussed earlier, in which badly behaved BE traffic is classified as LBE. This design, of course, has both benefits and limitations. The benefits are that less complex RED profiles are needed for the queue and that the buffer for BE in-contract traffic can use the entire length of the buffer. The limitations are that more queues are needed and thus the available buffer space is shared with more queues and that traffic can be severely reordered and jitter increased compare with having the traffic in the same physical buffer, resulting in more retransmissions. Current end-user TCP stacks can reorder packets into the proper order without much impact on resources, but this is not possible for UDP packets. So, applying RED profiles to UDP is a poor design.

8.7   Head versus Tail RED

While the specifics of the RED operation depends on the platform and the vendor, there are two general types that are worth describing: head and tail RED. The difference between them is where the RED operation happens, at the head of the queue or at the tail. As with most things, there are upsides and downsides to both implementations. Before discussing this, it is important to note that head and tail RED have nothing to do with queue tail drops or packet aging (queue head drops) described in earlier chapters. RED drops occur independently of any queue drops because either the buffer is full or packets are about to be aged out.

With head RED, all packets are queued, and when a packet reaches the head of the queue, a decision is made about whether to drop it. The downside is that packets dropped by head RED are placed in the queue and travel through it and are dropped only when they reach the head of the queue.

Tail RED exhibits the opposite behavior. The packet is inspected at the tail of the queue, where the decision to drop it is made. Packets marked by tail RED

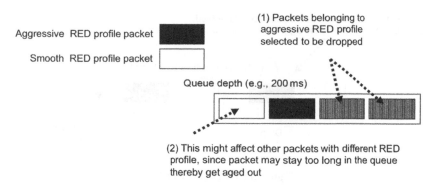

Figure 8.13   Head-based RED

are never queued. The implementation of the RED algorithm for both head and tail RED is not very different. Both monitor the queue depth, and both decide whether to drop packets and at what rate to drop them. The difference between the two is how they behave.

Head-based RED is good for using high-speed links efficiently because RED is applied to packets in the queue. Utilization of the link and its buffer arc thereby maximized, and short bursts can be handled easily without the possibility of dropping packets. On low-speed links, head-based RED has drawbacks because packets that are selected to be dropped are in the way of packets that are not being dropped. This head-of-line blocking situation means that packets not selected to be dropped must stay in the queue longer, a situation that might affect the available queue depth and might result in unintended packets being dropped. To avoid this situation, the RED profiles in the same queue need to be considerably different to effect more aggressive behavior (see Figure 8.13).

On the other hand, tail-based RED monitors the queue depth and calculates the average queue length of the stream inbound to the queue. Tail-based RED is good for low-speed links because it avoids the head-of-line blocking situation. However, packet dropping can be too aggressive, with the result that the link and buffer might not be fully utilized. To avoid RED being too aggressive, profiles must be configured to allow something of a burst before starting to drop packets (see Figure 8.14).

One more academic reason also favors head-based RED over tail-based RED: in congestion scenarios, the destination is made aware of the problem sooner, following the maxim that bad news should travel fast. Consider the scenario in Figure 8.15.

In Figure 8.15, assume congestion is occurring and packets need to be dropped. Head RED operates at the queue head, so it drops P1, immediately making the

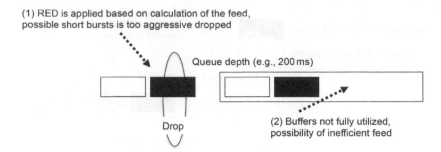

(1) RED is applied based on calculation of the feed,
possible short bursts is too aggressive dropped

Queue depth (e.g., 200 ms)

Drop

(2) Buffers not fully utilized,
possibility of inefficient feed

Figure 8.14 Tail-based RED

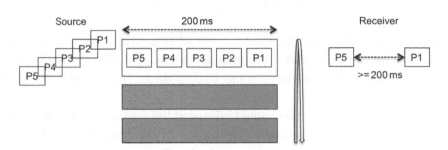

Source   200 ms   Receiver

| P5 | P4 | P3 | P2 | P1 |

P1

>= 200 ms

Figure 8.15 Bad news should travel fast

destination aware of the congestion because P1 never arrives. Tail RED operates
at the queue's tail, so when congestion occurs, it stops packet P5 from being
placed in the queue and drops it. So the destination receives P1, P2, P3, and P4,
and only when P5 is missed is the destination made aware of congestion.
Assuming a queue length of 200 ms as in this example, the difference is that the
destination is made aware of congestion 200 ms sooner with head RED. For
modern TCP-based traffic with more intelligent retransmission features, this
problem is more academic in nature. The receiver triggers, for example, SACK
feedback in the ACK messages to inform the sender about the missing segment.
This makes TCP implementations less sensitive to segment reordering.

8.8 Segmented and Interpolated RED Profiles

A RED profile consists of a list of pairs, where each pair consists of a queue fill
level and the associated drop probability. On an X–Y axis, each pair is represented by a single point. Segmented and interpolated RED are two different
methods for connecting the dots (see Figure 8.16). However, both share the
same basic principle of operation: first, calculate an average queue length, and
then generate a drop probability factor.

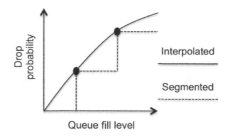

Figure 8.16   Interpolated and segmented RED profiles

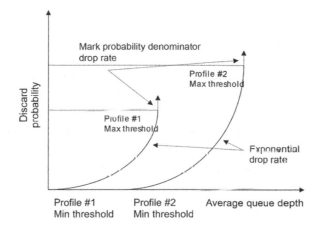

Figure 8.17   Interpolated RED drop curve example 1

Segmented RED profiles assume that the drop probability remains unchanged from the fill level at which it was defined, right up to the fill level at which a different drop probability is specified. The result is a stepped behavior.

The design of interpolated RED profiles is based on exponential weight to dynamically create a drop probability curve. Because implementations of RED are, like most things with QOS, vendor specific, different configurations are needed to arrive at the same drop probability X based on the average queue length Y.

One implementation example creates the drop probability based on the input of a minimum and a maximum threshold. Once the maximum threshold is reached, the tail-drop rate is defined by a probability weight. Thus, the drop curve is generated based on the average queue length (see Figure 8.17).

A variation of this implementation is to replace the weight with averages that are computed as percentages of discards or drops compared with average queue lengths (see Figure 8.18).

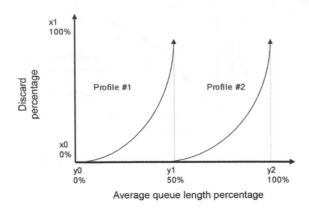

Figure 8.18   Interpolated RED drop curve example 2

What is the best approach? Well, it is all about being able to generate a configuration that performs with predicted behavior while at same time being flexible enough to change in case new demands arise.

Extrapolating this theoretical exercise to real life, it is the case that most implementations have restrictions. Even if you have an interpolated design with generated exponential length and drop values, there are limitations on the ranges of measured points. For example, there is always a limit on the number of status ranges of elements or dots, 8, 16, 32, or 64, depending on the vendor implementation and hardware capabilities. Also, there are always rounding and averages involved in the delta status used to generate the discards as compared with the queue length ratio.

In a nutshell, there are limited proofs to show any differences between segmented and interpolate implementations. In the end, what is more important is to be able to configure several profiles that apply to different traffic control scenarios and that can be implemented on several queues or on the same queue.

8.9   Conclusion

So, does RED work in today's networks? The original version of RED was designed with an eye on long-lasting TCP sessions, for example, file transfers. By dropping packets randomly, RED would provide hosts with the opportunity to lower the CWND when they received duplicate ACKs. With the explosion of web surfing, there has been an increase in the number of short-lived sessions. RED is not very efficient for these short web transactions, because a web session often just resends packets after a timeout as congestion mechanism. But also, as discussed in Chapter 4, with the recent file sharing and downloading of streamed media, long-lasting sessions have returned to some extent. So is RED useful now? The answer is probably yes, but a weak yes.

The value of using RED profiles for TCP flows is questionable in environments that use the latest TCP implementations, in which the TCP stacks are able to take a considerable beating, as described in Chapter 4. Features such as SACK, which in each ACK message can provide pointers to specific lost segments, help TCP sessions maintain a high rate of delivery, thus greatly reducing the need for massive amounts of retransmission if a single segment arrives out of order. The result is that any RED drop levels need to be aggressive to slow down a single host; that is, dropping one segment is not enough.

Because RED is not aware of sessions, it cannot punish individual "bad guys" that ramp up the CWND and rate and that take bandwidth from others. The whole design idea behind RED is that many packets need to be in the queue belonging to certain sessions, thereby increasing the chance of dropping packets for these bandwidth-greedy sessions.

However, RED is still a very powerful tool for differentiating between different traffic types inside the queue itself, for example, if packets have different DSCP markings.

Reference

[1] Floyd, S. and Jacobson, V. (1993) *Random Early Detection Gateways for Congestion Avoidance.* Berkeley, CA: Lawrence Berkeley Laboratory, University of California.

Further Reading

Borden, M. and Firoiu, V. (2000) A Study of Active Queue Management for Congestion Control, IEEE Infocom 2000, Tel Aviv-Yafo, Israel.

Croll, A. and Packman, E. (2000) *Managing Bandwidth: Deploy QOS in Enterprise Networks.* Upper Saddle River, NJ: Prentice Hall.

Goralski, W. (2010) JUNOS Internet Software Configuration Guide: "Class of Service Configuration Guide," Release 10.1, Published 2010. www.juniper.net (accessed August 18, 2015).

Heinanen, J., Baker, F., Weiss, W. and Wroclawski, J. (1999) RFC 2597, Assured Forwarding PHB Group, June 1999. https://tools.ietf.org/html/rfc2597 (accessed August 18, 2015).

Jacobson, V., Nichols, K. and Poduri, K. (1999) RFC 2598, An Expedited Forwarding PHB, June 1999. https://tools.ietf.org/html/rfc2598 (accessed August 18, 2015).

Nichols, K., Blake, S., Baker, F. and Black, D. (1998) RFC 2474, Definition of the Differentiated Services Field, December 1998. https://tools.ietf.org/html/rfc2474 (accessed August 18, 2015).

Postel, J. (1981) RFC 793, Transmission Control Protocol—DARPA Internet Protocol Specification, September 1981. https://tools.ietf.org/rfc/rfc793.txt (accessed August 18, 2015).

Semeria, C. (2002) *Supporting Differentiated Service Classes: Active Queue Memory Management*, White Paper. Sunnyvale, CA: Juniper Networks, Inc.

Stevens, R. (1994) *TCP Illustrated, Vol. 1: The Protocols.* Upper Saddle River, NJ: Addison-Wesley.

Zang, H. (1998) *Models and Algorithms for Hierarchical Resource Management.* Pittsburgh, PA: School of Computer Science, Carnegie Mellon University.

Part III

Case Studies

9

The VPLS Case Study

In the previous chapters of this book, the main focus has been the analysis of the QOS tools, one by one, and in Chapter 3 we focused on the challenges of a QOS deployment. We have now introduced all the different parts of the QOS equation and are ready to move forward and present three case studies that illustrate end-to-end QOS deployments in the format of case studies. The case study in this chapter focuses on a Virtual Private LAN Service (VPLS) scenario, the second one is focused in the internals of a Data Center, and the third one in a mobile network. The reason for selecting these three realms is due to the fact that they are the most challenging ones at present, but unquestionably the lessons learned from it can be translated to other less complex realms, like an enterprise deployment by just removing the MPLS component out of the equation of this case study. Using these case studies, we glue together the QOS concepts and tools we have discussed and also explain how to cope with some of the challenges highlighted in Chapter 3. However, as previously discussed, any QOS deployment is never completely independent of other network components, such as the network routing characteristics, because, for example, routing can place constraints on the number of hops, or which particular hops, are crossed by the traffic. So it is not accurate to present a QOS case study without first analyzing the network and its characteristics. However, discussion of the other network components is limited only to what is strictly necessary so that the focus of the case studies always remains on QOS.

This chapter starts by presenting an overview of the network technologies being used and how the particular services are implemented. Then we define

---

QOS-Enabled Networks: Tools and Foundations, Second Edition. Miguel Barreiros and Peter Lundqvist.
© 2016 John Wiley & Sons, Ltd. Published 2016 by John Wiley & Sons, Ltd.

the service specifications, starting off with some loose constraints that we discuss and tune as the case study evolves. Finally, we discuss which QOS tools should be used and how they should be tuned to achieve the desired end result. This is illustrated by following a packet as it traverses across the network.

9.1   High-Level Case Study Overview

Before discussing the specific details of this case study, it will be helpful to provide a high-level perspective of its design and components. This case study focuses on a service provider network that offers VPN services to its customers. For several years, VPN services have provided the highest revenue for service providers, and when it comes to Layer 2 services, VPLS is one of the most popular ones being commonly used, not just by the service providers but also for Data Center interconnections. For a service provider network, one primary reason is that VPLS is implemented with MPLS tunnels, also referred to as label-switched paths (LSPs), as illustrated in Figure 9.1.

These tunnels carry the VPLS traffic (shown as white packets in Figure 9.1) between the routers on which the service end points are located. So the QOS deployment interacts with MPLS, and so must be tailored to take advantage of the range of MPLS features. Control traffic, represented in Figure 9.1 as black triangles, is what keeps the network alive and breathing. This traffic is crucial, because without it, for example, the MPLS tunnels used by customer traffic to travel between the service end points cannot be established. VPLS is the support used for the services.

In this case study, the customer has different classes of service from which to choose. Here, the network has four classes of service: real time (RT), business (BU), data (DA), and best effort (BE). We explain later in this chapter how these classes differ. Looking at traffic flow, when customer traffic arrives at the network, it is split into a maximum of four classes of service, as illustrated in Figure 9.2.

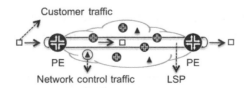

Figure 9.1   MPLS tunnels used to forward the customer traffic

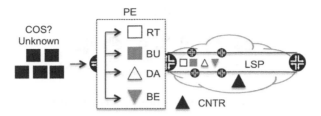

Figure 9.2   Customer traffic

From the network perspective, traffic arriving from the customer has no class of service assigned to it, so the first task performed by the edge router is to split the customer traffic into the different classes of service. This split is made according to the agreement established between the customer and the service provider network at the service end point. Then, based on the service connectivity rules, traffic is placed inside one or more LSPs toward the remote service end points.

Figure 9.2 shows two major groups of traffic inside the network, customer and control traffic, and the customer traffic can be split into four subgroups: RT, BU, DA, and BE. A key difference between customer and control traffic is that the customer traffic transits the network, that is, it has a source and destination external to the network, while control traffic exists only within the network.

This overview of the network internals and the service offered is our starting point. As the case study evolves, we will introduce different technologies and different types of traffic. Let us now focus on each network component individually.

9.2   Virtual Private Networks

In a nutshell, virtual private networks (VPNs) provide connectivity to customers that have sites geographically spread across one or more cities, or even countries, sparing the customers from having to invest in their own infrastructure to establish connectivity between its sites.

However, this connectivity carries the "Private" keyword attached to it, so the customer traffic must be kept private. This has several implications for routing, security, and other network technologies. For QOS, this requirement implies that the customer traffic must be delivered between sites based on a set of parameters defined and agreed upon by both parties, independently of any other traffic that may be present inside the service provider's network. This set of parameters is usually defined in a service-level agreement (SLA). To meet the SLA, the customer traffic exchanged between its several sites is placed into one or more

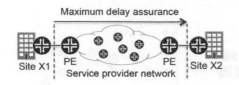

Figure 9.3 Delay assurance across a service provider network

classes of service, and for each one, a set of parameters is defined that must be met. For example, one parameter may be the maximum delay inserted into the transmission of traffic as it travels from customer site X1 to X2, an example illustrated in Figure 9.3.

The VPN service has two variants, Layer 3 (L3) and Layer 2 (L2). To a certain degree, the popularity of MPLS as a technology owes much to the simplicity it offers for service providers implementing VPNs. The success of L2VPNs is closely linked to the Ethernet boom as a transport technology (because its cost is lower than other transmission mediums) and also to the fact that L2VPNs do not need a routing protocol running between the customer device and the service provider's PE router. So the PE router does not play a role in the routing of customer traffic. One result is that the customer device can be a switch or a server that does not necessarily support routing functionality.

9.3 Service Overview

The service implemented using VPLS operates as a hub-and-spoke VLAN, and the network customer-facing interfaces are all Ethernet.

A hub site can be seen as a central aggregation or traffic distribution point, and spoke sites as transmitters (or receivers) toward the hub site, as illustrated in Figure 9.4.

In terms of connectivity, a hub site can communicate with multiple spoke sites or with other hub sites if they exist, but a spoke site can communicate only with a single hub site. These connectivity rules mean that for Layer 2 forwarding, there is no need for MAC learning or resolution at the spoke sites, because all that traffic that is transmitted (or received) has only one possible destination (or source), the hub site.

There are several applicable scenarios for this type of service. One example is a hub site that contains multiple financial services servers that wish to distribute information to multiple customers in wholesale fashion, meaning that the customers have no connectivity with each other. Another example is Digital Subscriber Line (DSL) traffic aggregation, in which multiple Digital Subscriber

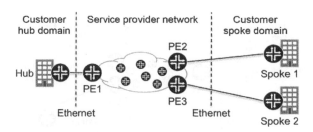

Figure 9.4 Hub-and-spoke VLAN

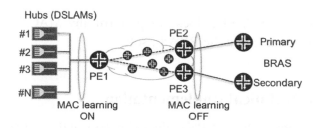

Figure 9.5 DSL scenario for the hub-and-spoke VLAN

Line Access Multiplexers (DSLAMs) aggregating DSL traffic connect to the hub points, and the hub-and-spoke VLAN carry that traffic toward a BRAS router located at a spoke point. This scenario can be expanded further by including a backup BRAS router at a secondary spoke point, to act, in case the primary router fails, as illustrated in Figure 9.5.

Hub points may or may not be able to communicate directly with each other. Typically, inter-DSLAM communication must be done via the BRAS router to apply accounting tools and to enforce connectivity rules. This means that "local switching" between hubs may need to be blocked.

The fact that no MAC learning is done at the spoke sites is a benefit for scaling, but it raises the question of how the BRAS router can select a specific DSLAM to which to send traffic. Typically, DSLAM selection is achieved by configuring different Ethernet VLAN tags on the hub interfaces facing the DSLAMs. For example, the Ethernet interface (either logical or physical) that faces DSLAM #3 at the hub point can be configured with an outer VLAN tag of 100 and an inner tag of 300, as illustrated in Figure 9.6.

If the primary BRAS wants to send traffic toward DSLAM #3, it must place these tags in the traffic it sends toward PE2, where the spoke point is located. PE2 forwards that traffic blindly (without any MAC address resolution) toward PE1, where the hub point is located. Then inside the hub point, the traffic is forwarded toward the Ethernet interface to which DSLAM #3 is connected.

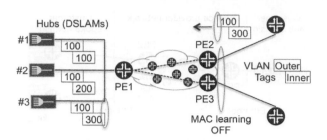

Figure 9.6   Packet forwarding inside the hub-and-spoke VLAN

As a side note, although PE2 does not perform any MAC learning, MAC learning must be enabled on the BRAS itself.

9.4   Service Technical Implementation

The main technology in the L2VPN realm to implement the service described in the previous section is VPLS, in particular using its mesh-group functionality combined with Layer 2 Circuits (l2ckt).

Within a VPLS domain, the PE routers must be fully meshed at the data plane so a forwarding split-horizon rule can be applied. This rule means that if a PE router receives traffic from any other PE, it can forward it only to customer-facing inter-faces but never to other PE routers. The VPLS protocol can be seen as transforming the network into one large and single Ethernet segment, so the existence of the forwarding split-horizon rule is necessary to ensure that there are no Layer 2 loops inside the network. This rule is the VPLS alternative to the Spanning Tree Protocols (STPs), which have been shown not to scale outside the LAN realm. The price to pay for the requirement of this full mesh between PE routers at the data plane is obviously scaling and complexity.

The mesh-group functionality breaks the VPLS behavior as a single Ethernet segment by dividing the VPLS domain into different mesh groups, where the obligation for a full mesh at the data plane between PE routers exists only within each mesh group. So traffic received from another PE in the same mesh group can indeed be forwarded to other PE routers, but only if they belong to a differ-ent mesh group. As a concept, mesh groups are not entirely new in the net-working world, because they already exist in the routing realm.

Let us now demonstrate how this functionality can be used to provide the connectivity requirements desired for the hub-and-spoke VLAN.

As illustrated in Figure 9.7, a VPLS instance is created at PE1 containing two mesh groups, one name MESH-HUB and other MESH-SPOKES. The

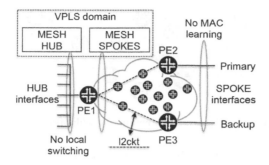

Figure 9.7 VPLS mesh groups

MESH-HUB group contains all the hub interfaces, and local switching between them can be blocked if required. The second mesh group, MESH-SPOKES, contains the spoke sites. It is from this mesh group that the two l2ckts, which terminate at PE2 and PE3, are created. Terminating the l2ckts inside the VPLS domain is the secret to not requiring any MAC learning at the spoke sites. L2ckts are often called pseudowires, because they emulate a wire across the network in the sense that any traffic injected at one end of the wire is received at the other end, so there is no need for any MAC resolution.

The topology shown in Figure 9.7 ensures resiliency since primary and secondary sites can be defined and also has the scaling gain that the spoke interfaces do not require any active MAC learning.

9.5   Network Internals

The previous sections have focused on the events that take place at the network edge. Now let us now turn our attention to the core of the network.

For routing, the network uses MPLS-TE with strict Explicit Routing Object (ERO), which allows 100% control over the path that traffic takes across the network, as illustrated in Figure 9.8.

Two distinct paths, a primary and a secondary, are always created between any two PE routers for redundancy. If a failure occurs on the primary path, the ingress node moves traffic onto the secondary path. However, to make things interesting, let us assume that each primary path crosses a maximum of four hops between the source and the destination, but for the secondary path that number increases to five. The importance of this difference with regard to QOS will become clearer later in this chapter, when we discuss the dimensioning of queue sizes.

While these sections have provided a high-level perspective of the network and services, a complete understanding requires additional knowledge about the

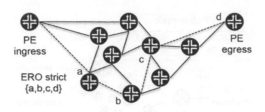

Figure 9.8   Network internal routing with MPLS-TE ERO

MPLS, VPLS, and L2VPN realms. A reader seeking more information about these topics should refer to the links provided in the Further Reading section of this chapter. However, to discuss the QOS part of this case study, it is not crucial to understand all the VPLS mechanics but just simply to accept the connectivity rules mentioned earlier and the values of the maximum number of hops crossed by the primary and secondary paths.

Now that the service and networks basics have been presented, we can move on to the QOS section of this case study.

9.6   Classes of Service and Queue Mapping

A customer accesses the network by a VLAN that is subject to an aggregate rate, which establishes the maximum ingress and egress rates of customer traffic entering and leaving the network. Once the rate is defined, the customer divides it into four classes of service: real time (RT), business (BU), data (DA), and best effort (BE), whose characteristics are summarized in Table 9.1.

The RT class of service is the most sensitive in terms of requirements, including an assurance that no more than 75 ms of delay be inserted into this traffic as it crosses the network between the service end points. As the reader can notice, the definition of the RT class of service is highly sensitive to jitter, but a maximum value is not defined. We discuss this more later in this case study.

The BU class of service is the second most sensitive in terms of requirements, although it is not as sensitive to delay and jitter as RT. However, it still has strict requirements regarding packet loss.

For the CIR/PIR model, the main interest lies in the DA class of service, because DA traffic will be split into green and yellow depending on its arrival rate.

In terms of admission control, an aggregate rate is applied to customer traffic as a whole as it enters the network, and then each class of service is subject to a specific rate. The only exception is BE traffic because this class has no CIR value, so no specific rate is applied to it. This means that while BE traffic has no

Table 9.1   Classes of service requirements

COS	Traffic sensitivity			CIR/PIR	Admission control
	Delay	Jitter	Packet loss		
RT	Max 75 ms	High	High	Only CIR	
BU	Medium	Medium	High	Only CIR	Y
DA		Low		Both	
BE				PIR	N

Table 9.2   Classes of service-to-queue mapping

COS	Queue	
	Priority	Number
RT	High	4
Internal	Medium	3 (Reserved)
BU	Medium	2
DA	Low	1
BE	Low	0

guarantees, if resources are not being used by traffic belonging to other classes of service, BE traffic can use them.

Based on the conclusions we drew in Chapter 3, we use a 1:1 mapping between queues and classes of service, as shown in Table 9.2. According to Table 9.2, we are selecting a priority level of high, rather than strict high, for the RT queue, although as we discussed in Chapter 7, a strict-high level assures a better latency. The reason for doing this is that implementation of a strict-high priority queue is heavily dependent on the router brand, while the operation of a high-priority queue on the PB-DWRR algorithm is defined in the algorithm itself. So to maintain the goal of a vendor-agnostic approach, which we set at the beginning of the book, we use a high-priority queue.

Also from Table 9.2, we see that although both DA green and yellow traffic exist, they are both mapped to queue number 1 (Q1) to avoid any packet reordering issues at the egress. Another point to note is that Q3 is reserved, because this is where the network control traffic—the traffic that keeps the network alive and breathing—will be placed. We set the Q3 priority to medium, which is the same as the BU priority, but lower than RT. This raises the interesting question that if control (CNTR) traffic is crucial to the network operation, why does any customer traffic have a higher priority at the queuing stage? This happens because although CNTR traffic is indeed crucial, it typically is less demanding in terms of delay and jitter than real-time traffic. Hence, setting it as medium priority should be enough for the demands of the CNTR traffic.

9.7   Classification and Trust Borders

In terms of classification, interfaces can be divided into two groups according to what is present at the other end, either another network router or a customer device.

When traffic arrives at the network from the customer (on a customer-facing interface), it is classified based on the User Priority bits present in the VLAN-tagged Ethernet frames, as illustrated in Figure 9.9.

Traffic arriving with a marking not associated with any class of service (represented as the 1XX combination, where X can be 0 or 1) is placed into the BE class of service, the lowest possible, as shown in Figure 9.9. Another option is to simply discard it.

Let us now describe how the classification process works on core-facing interfaces.

Between network routers, customer traffic is forwarded inside MPLS tunnels and control traffic travels as native IP, so the classifier needs to inspect the EXP markings of the MPLS packets and the DSCP marking of the IP packets. This behavior is illustrated in Figure 9.10.

We choose a DSCP value of 100 000 (it could be any value) to identify control traffic inside the network. It is the responsibility of any network router to place this value in the DSCP field of the control traffic it generates. Using this classifier guarantees that other routers properly classify control traffic and place it in the correct egress queue. Another valid option is to use the IP Precedence field of the IP Packets instead of the DSCP.

As also shown in Figure 9.10, two distinct EXP markings are used for the class of service DA. This is required to ensure that the differentiation made by

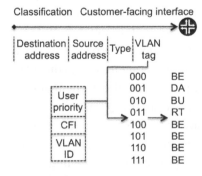

Figure 9.9   Classification at customer-facing interfaces

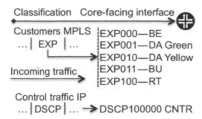

Classification   Core-facing interface

Customers MPLS   EXP000—BE
...| EXP |...   EXP001—DA Green
                      EXP010—DA Yellow
Incoming traffic   EXP011—BU
                      EXP100—RT

Control traffic IP
...|DSCP|... → DSCP100000 CNTR

Figure 9.10   Classification at core-facing interfaces

Table 9.3   Classification rules

Classification	Interfaces		
	Customer facing	Core facing	
COS	User priority	EXP	DSCP
BE	000	000	
DA	001	001 (Green)	
		010 (Yellow)	
BU	010	011	
CNTR			100 000
RT	011	100	
Not used	1XX	101110111	

the ingress PE when implementing the CIR/PIR model, which colors DA traffic as green or yellow, can be propagated across all other network routers.

These classification rules are summarized in Table 9.3.

Consistency is achieved by ensuring that all the classifiers present on the network routers, either on customer- or core-facing interfaces, comply with the rules in Table 9.3. These rules ensure that there is never any room for ambiguities regarding the treatment to be applied to any given packet. However, how can we ensure that the packets' QOS markings received at any given classifier are correct, for example, that a packet received on a core-facing interface with an EXP marking of 011 is indeed a BU packet? This is achieved by admission control and classification on customer-facing interfaces and by rewrite rules on core-facing interfaces, topics detailed later in this chapter.

9.8   Admission Control

Before traffic enters the network, it is crucial that it complies with the agreement established between both parties regarding the hired traffic rates.

As previously mentioned, the aggregate rate represents the total amount of bandwidth hired by the customer at a specific access. Inside that aggregate rate are particular rates defined for the classes of service RT, BU, and DA. (The DA class of service is a special case, because two rates are defined for it, both a CIR and a PIR as shown in Table 9.1.) The BE class of service is not subject to any specific rate. Rather, it uses any leftovers generated when there is no traffic from other classes of service.

The first part of the admission control is established by a policer implemented in a hierarchical fashion, first, by making the traffic as a whole comply with the aggregate rate, and then applying the specific contracted rate to each class of service except BE.

This policer, together with the classification tool, achieves the desired behavior for ingress traffic received on customer-facing interfaces. Traffic that exceeds the hired rate is dropped, traffic with an unknown marking is placed in the BE class of service, and traffic in the DA traffic class is colored green or yellow according to its arrival rate.

No policing is needed at any other points of the network, because all customer traffic should be controlled before passing the customer-facing interface and thus before entering the service provider network. Also, no policing is applied to CNTR traffic.

The definition of the rates at which the traffic is policed is a direct consequence of established agreements, so they can vary according to each customer's requirements. The exception, though, is dimensioning the burst size limit parameter. Each router vendor has recommended values for the burst size limit parameter on their platforms. However, as with many things inside the QOS realm, choosing a value for the burst size limit parameter is not an exact science. Recalling the discussion in Chapter 6, the strategy is to start with a small value, but then change it, based on feedback received from customers and from monitoring the interfaces on which policing is applied.

9.9   Rewrite Rules

Rewrite rules are necessary to ensure that all MPLS traffic that leaves a network router, either a PE or a P router, has the EXP markings specified in Table 9.3. The more interesting scenario is that of the PE router, on which traffic arrives from a customer-facing interface on an Ethernet VLAN (in the sense that the classifier inspects the User Priority field in the header) and then leaves encapsulated inside an MPLS tunnel toward the next downstream network router, as illustrated in Figure 9.11.

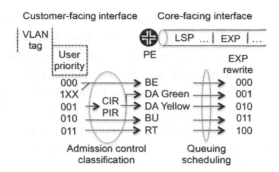

Figure 9.11   Applicability of EXP rewrite rules

Figure 9.11 shows that after admission control and classification are applied, traffic is split into different classes of service, so it is on this router the classification task is achieved. Several other QOS tools are then applied to the traffic before it exits this router toward the next downstream neighbor. When the neighbor receives the customer traffic, which is encapsulated inside an LSP, it classifies it based on its EXP markings.

The key role of rewrite rules is to enforce consistency, to ensure that the EXP markings of the traffic exiting a core-facing interface comply with the classification rules used by the next downstream router, as defined in Table 9.3. For example, if a packet is classified by a router as BU, the rewrite rules ensure that it leaves with an EXP marking of 011, because for the classifier on the next downstream router, a packet with a marking of 011 is classified into the BU class of service.

Returning to Figure 9.11, we see that the forwarding class DA, which is identified by a User Priority value of 001 when it arrives from the customer at the PE router, is a special case. As specified in Table 9.1, the metering tool on the router applies a CIR/PIR model, which creates two colors of DA traffic, green and yellow.

The EXP rewrite rule maintains consistency by ensuring that DA green and DA yellow traffic depart from the core-facing interface with different EXP markings. These markings allow the next downstream router to differentiate between the two, using the classification rules defined in Table 9.3, which associates an EXP marking of 001 with DA green and 010 with DA yellow. Once again, we see the PHB model in action.

The above paragraphs describe the required usage of rewrite rules in core-facing interfaces. Let us now evaluate their applicability in customer-facing interfaces.

As illustrated in Figure 9.9, packets arriving from the customer that have a marking of 1XX are outside the agreement established between both parties, as specified by Table 9.3. As a consequence, this traffic is placed in the BE class of service. From an internal network perspective, such markings are irrelevant, because all customer traffic is encapsulated and travels through in MPLS tunnels, so all QOS decisions are made based on the EXP fields and the EXP classifiers and rewrite rules ensure there is no room for ambiguity. However, one problem may occur when traffic is desencapsulated at the remote customer-facing interface, as illustrated in Figure 9.12.

In Figure 9.12, traffic arrives at the ingress PE interface with a marking of 1XX, so it is placed into the BE class of service and is treated as such inside the network. However, when the traffic is delivered to the remote customer site, it still possesses the same marking. It can be argued that the marking is incorrect because it is outside what has been established in Table 9.3.

To illustrate the applicability of rewrite rules for customer-facing interfaces, we apply a behavior commonly named as bleaching, as illustrated in Figure 9.13. Figure 9.13 shows that traffic that has arrived at the ingress PE with a QOS marking outside the agreed range is delivered to the remote service end point with that QOS marking rewritten to zero. That is, the QOS marking has been bleached. As with all QOS tools, bleaching should be used when it helps achieve a required goal.

As previously discussed in Chapter 2, QOS tools are always applied with a sense of directionality. As such, it should be pointed out that the rewrite rules described above are applied only to traffic exiting from an interface perspective. They are not applied to incoming traffic on the interface.

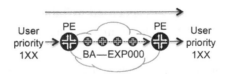

Figure 9.12 Customer traffic with QOS markings outside the specified range

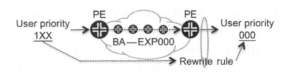

Figure 9.13 Bleaching the QOS markings of customer traffic

9.10   Absorbing Traffic Bursts at the Egress

The logical topology on top of which the service is supported is a hub and spoke, so traffic bursts at the egress on customer-facing interfaces are a possibility that need to be accounted for. As such, the shaping tool is required. However, as previously discussed, the benefits delivered by shaping in terms of traffic absorption come with the drawback of introducing delay.

The different classes of service have different sensitivities to delay, which invalidates the one-size-fits-all approach. As a side note, shaping all the traffic together to the aggregate rate is doable, but it would have to be done with a very small buffer to ensure that RT traffic is not affected by delay. Using such a small buffer implies a minimal burst absorption capability, which somewhat reduces the potential value of the shaping tool.

So, to increase granularity, the shaping tool is applied in a hierarchical manner, split into levels. The first level of shaping is applied to traffic that is less sensitive to delay, specifically, BE, DA, and BU. The idea is to get this traffic under control by eliminating any bursts. After these bursts have been eliminated, we move to the second level, applying shaping to the output of the first level plus to the RT traffic. This behavior is illustrated in Figure 9.14.

The advantage of applying several levels of shaping is to gain control over the delay buffer present at each level. Having a large buffer for the classes of service BE, DA, and BU can be accomplished without compromising the delay that is inserted into the RT class of service.

In terms of shaping, we are treating BU traffic as being equal to BE and DA even though, according to the requirements defined in Table 9.1, they do not have equal delay sensitivities. Such differentiation is enforced at the queuing and scheduling stage by assigning a greater priority to the queue in which BU traffic is placed.

9.11   Queues and Scheduling at Core-Facing Interfaces

The routing model so far is that all the traffic sourced from PE X to PE Y travels inside a single LSP established between the two, and the maximum number of hops crossed is four, which includes both the source and destination PE routers.

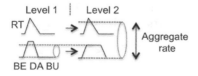

Figure 9.14   Hierarchical shaper

If the primary path fails, the secondary path is used, which raises the maximum number of hops crossed to five. Later in this case study, we analyze the pros and cons of having multiple LSPs with bandwidth reservations between two PE routers.

In terms of granularity, a core-facing interface is only class of service aware, so all customer traffic that belongs to a certain class of service is queued together, because it all requires the same behavior. Also, control traffic on the core interface needs to be accounted for.

The queuing mechanism chosen is PB-DWRR because of the benefits highlighted in Chapter 7. How a queuing and scheduling mechanism operates depends on three parameters associated with each queue: the transmit rate, the length, and its priority. We have already specified the priority to be assigned to each queue in Table 9.2 and have explained the reasons for these values, so we now focus on the other two parameters.

To handle resource competition, an interface has a maximum bandwidth value associated with it, which can either be a physical value or artificially limited, and a total amount of buffer. Both of these are divided across the queues that are present on that interface, as illustrated in Figure 9.15.

Choosing a queue transmit rate depends on the expected amount of traffic of the class of service that is mapped to that specific queue, so it is a parameter tied to network planning and growth. Such dimensioning usually also takes into account possible failure scenarios, so that if there is a failure, some traffic is rerouted and ideally other links have enough free bandwidth to cope with the rerouted traffic. (We return to this topic when we discuss multiple LSP with bandwidth reservations later in this chapter.) However, it is becoming clearer that there is always a close tie between a QOS deployment and other active processes in the network, such as routing or the link bandwidth planning.

Regarding the amount of buffer allocated to each queue, the RT class of service is special because of its rigid delay constraint of a maximum of 75 ms. When RT traffic crosses a router, the maximum delay inserted equals the queue length in which that traffic is placed. So the value for the queue length to be configured at each node can be determined by dividing the 75 ms value by the

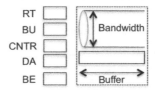

Figure 9.15 Interface bandwidth and buffer division across the classes of service

maximum number of hops crossed. However, such division should take into account not just the scenario of the traffic using the primary path, which has four hops, but also possible failure scenarios that cause the traffic to use the secondary path, which has five hops. Not accounting for failure of the primary path may cause traffic using the secondary path to violate the maximum delay constraint. So, dividing the maximum delay constraint of 75 ms by five gives an RT queue length of 15 ms.

This dimensioning rule applies to both core- and customer-facing interfaces, because the maximum number of hops crossed also needs to take into account the egress PE interface, where traffic is delivered to the customer. Also the RT queue is rate limited, as explained in Chapter 7, to keep control over the delay inserted.

After the queue length for RT has been established to comply with the delay requirement, an interesting question is what assurances such dimensioning can offer regarding the maximum jitter value. From a purely mathematical point of view, if the delay varies between 0 and 75 ms, the maximum delay variation is 75 ms. However, the jitter phenomenon is far more complex than this.

When two consecutive packets from the same flow are queued, two factors affect the value of jitter that is introduced. First, the queue fill level can vary, which means that each packet reaches the queue head in a different amount of time. Second, there are multiple queues and the scheduler is jumping between them. Each time the scheduler jumps from serving the RT queue to serving the other queues, the speed of removal from the queue varies, which introduces jitter. Obviously, the impact of such scheduler jumps is minimized as the number of queues decreases. Another approach to reduce jitter is to give the RT queue a huge weight in terms of the PW-DWRR algorithm to ensure that in-contract traffic from other classes of service does not hold up the scheduler too often and for too much time. However, a balance must be found regarding how much the RT traffic should benefit at the expense of impacting the other traffic types.

So in a nutshell, there are tactics to minimize the presence of jitter, but making an accurate prediction requires input from the network operation, because the average queue fill levels depend on the network and its traffic patterns.

The guideline for the other queues is that the amount of interface buffer allocated to the queue length should be proportional to the value of the assigned transmit rate. So, for example, if the BU queue has a transmit rate of 30% of the interface bandwidth, the BU queue length should be equal to 30% of the total amount of buffer available on the interface. This logic can be applied to queues that carry customer traffic, as well as to queues that carry the network internal traffic such as the CNTR traffic, as illustrated in Figure 9.16. Here, x, x1, x2,

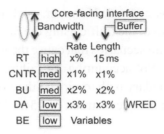

Figure 9.16   Queuing and scheduling at core-facing interfaces

and x3 are variables that are defined according to the amount of traffic of each class of service that is expected on the interface to which the queuing and scheduling are applied.

The DA queue has both green and yellow traffic, so a WRED profile is applied to protect the queuing resources for green traffic.

A special case is the BE queue, because its transmit rate increases and decreases over time, depending on the amount of bandwidth available when other classes of service are not transmitting. Because of the possible variations in its transmit rate, the BE queue length should be as high as possible to allow it to store the maximum amount of traffic. The obvious implication is a possible high delay value inserted into BE traffic. However, such a parameter is considered irrelevant for such types of traffic. Another valid option to consider is to use a variable queue length for BE, a process described in Chapter 8 as MAD, following the logic that if the transmit rate increases and decreases, the queue length should also be allowed to grow and shrink accordingly.

9.12   Queues and Scheduling at Customer-Facing Interfaces

Some of the dimensioning for queuing and scheduling at a customer-facing interface is the same as for core-facing interfaces. However, there are a few crucial differences.

In terms of granularity, customer-facing interfaces are aware both of the class of service and the customer. For example, on a physical interface with multiple hub-and-spoke VLANs, the queuing and scheduling need to be done on a per-VLAN basis, because each VLAN represents a different customer.

Another difference is the dimensioning of the transmit rates, because at this point the focus is not the expected traffic load of each class of service as it was on a core-facing interface, but rather the agreement established with the customer regarding how much traffic is hired for each class of service. For the bandwidth sharing, the value to be split across the configured queues is not the interface maximum bandwidth but is the value of the defined shaping rate. Because the customer-facing interface is the point at which traffic is delivered to the customer, a shaper is established to allow absorption of traffic bursts, as previously discussed in Chapter 6. In a certain way, the shaper can be seen as guarding the exit of the scheduler.

Also, a customer-facing interface does not handle any CNTR traffic, so Q3 is not present.

All other points previously discussed in the core-facing interfaces section, including queue priorities, queue length dimensioning, and applying WRED on the DA queue, are still valid.

9.13   Tracing a Packet through the Network

The previous sections of this chapter have presented all the QOS elements that are active in this case study. Now let us demonstrate how they are combined by tracing a packet across the network.

We follow the path taken by three packets named RT1, DA1, and DA2 and a fourth packet that arrives with an unknown marking in its User Priority field, which we call UNK. All these packets belong to a hub-and-spoke VLAN, and they cross the network inside a primary LSP established between two PE routers, as illustrated in Figure 9.17.

The four packets are transmitted by the customer and arrive at the ingress PE customer-facing interface, where they are subject to admission control and classification based on their arrival rate and the markings in their User Priority field.

At the ingress PE, we assume that packet RT1 arrives below the maximum rate agreed for RT traffic, so it is admitted into the RT class of service. For the DA packets, when DA2 arrives, it falls into the interval between CIR and PIR so it is colored yellow, and DA1 is colored green. Packet UNK arrives with a

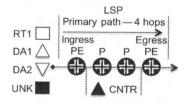

Figure 9.17   Tracing a packet across the network

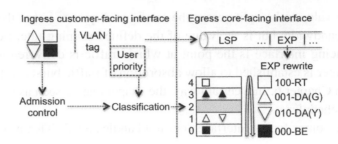

Figure 9.18   Packet walkthrough across the ingress PE router

marking of 111 in its User Priority field, and because this is outside the range established in Table 9.3, it is classified as BE. All four packets are now classified, so we can move to the core-facing egress interface.

The core-facing interface of the ingress PE router has five queues, four of which carry customer traffic belonging to this and other customers, and the last queue, Q3, carries the network's own control traffic, as illustrated in Figure 9.18, which again shows CNTR packets as black triangles. Also, we are assuming that there is no BU traffic, so Q2 is empty.

The RT packet is placed in Q4, both DA1 and DA2 are placed in Q1, and UNK is placed in Q0, following the class of service-to-queue mapping established in Table 9.2. How long it takes for the packets to leave their queues depends on the operation of the PB-DWRR algorithm, which considers the queue priorities and transmit rates.

For packet forwarding at this PE router, the customer traffic is encapsulated inside MPLS and placed inside an LSP, which crosses several P routers in the middle of the network and terminates at the egress PE.

An EXP rewrite rule is applied to ensure that traffic leaving this PE router has the correct EXP marking, as defined in Table 9.3, because those same EXP markings are the parameters evaluated by the ingress core-facing interface on the next P router crossed by the LSP.

At any of the P routers between the ingress and egress PE routers, the behavior is identical. At ingress, the classifier inspects the packets' EXP markings and assigns the packets to the proper classes of service. It should be pointed out that P routers understand the difference in terms of color between packets DA1 and DA2, because these packets arrive with different EXP values, another goal accomplished by using the EXP rewrite rule. The network control traffic is classified based on its DSCP field, as illustrated in Figure 9.19. The P router's egress interface needs to have queuing and scheduling and an EXP rewrite rule.

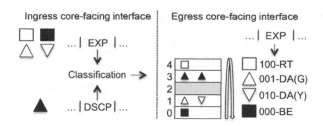

Figure 9.19   Packet walkthrough across the P router

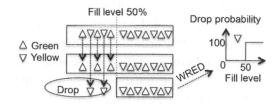

Figure 9.20   WRED operation

As previously mentioned, WRED is active on Q1 to protect queuing resources for green traffic. Let us now illustrate how that resources protection works by considering the scenario illustrated in Figure 9.20.

In Figure 9.20, when the fill level of Q1 goes above 50%, the drop probability for any newly arrived DA yellow packet (represented as inverted triangles) as established by the WRED profile is 100, meaning that they are all dropped. This implies that Q1 can be seen as having a greater length for green packets than for yellow packets. Following this logic, Q1 is effectively double the size for green packets than for yellow ones. So on any router crossed by packets DA1 and DA2, if the fill level of Q1 is more than 50%, DA2 is discarded. This packet drop should not be viewed negatively; it is the price to pay for complying with the CIR rate.

The UNK packet is placed in Q0, because it was considered as BE by the ingress PE, but it contains the same marking of 111 in the User Priority field that it had when it arrived at the ingress PE. This has no impact because all the QOS decisions are made based on the EXP field, so the value in the User Priority field of the packet is irrelevant.

The final network point to analyze is the egress PE router. At this router, traffic arrives at the ingress core-facing interface encapsulated as MPLS and is delivered to the customer on an Ethernet VLAN, as illustrated in Figure 9.21.

Because it is a customer-facing interface, there is no CNTR traffic, so Q3 is not present.

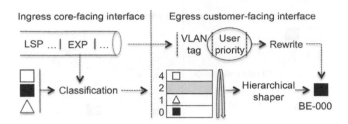

Figure 9.21 Packet passage through the egress PE router

Packet DA2 is not present at the egress PE router, because it was dropped by the WRED profile operation on one of the routers previously crossed.

When the customer traffic is delivered on an Ethernet VLAN, the markings present in the User Priority field once again become relevant. A rewrite rule on the customer-facing egress interface must be present to ensure that the User Priority field in the UNK packet is rewritten to zero. This step indicates to the customer network the fact that the ingress PE router placed this packet in the BE class of service.

The previous figures have shown packets arriving and departing simultaneously. This was done for ease of understanding. However, it is not what happens in reality because it ignores the scheduling operation, which benefits some queues at the expense of penalizing others. The time difference between packets arriving at an ingress interface depends on the PB-DWRR operation preformed at the previous egress interface. Because of the queue characteristics, we can assume that packet RT1 arrives first, followed by DA1, and only then does packet UNK arrive, and packet DA2 doesn't arrive at all because it was dropped by WRED previously.

Although both Q1 and Q0 have the same priority value of low, it is expected that packets in Q1 are dequeued faster, because this queue has an assured transmit rate. Q0 has a transmit rate that varies according to the amount of free bandwidth available when other queues are not transmitting.

9.14 Adding More Services

So far, this case study has considered a mapping of one type of service per each class of service, which has made the case study easier to follow. However, as discussed throughout this book, this is not a scalable solution. The key to a scalable QOS deployment is minimizing the behavior aggregates implemented inside the network and to have traffic belonging to several services reuse them.

So let us analyze the impact of adding a new service that we call L3, which is delivered over a Layer 3 VPN. We consider that its requirements are equal to those previously defined for the RT class of service. Implementing this new service requires a change in the technology used to provide connectivity between customer sites, which has several implications for network routing and VPN setup. However, the pertinent question is what of the QOS deployment presented so far needs to change.

From the perspective of a P router, the changes are very minimal. Packets belonging to RT and L3 arrive at the P router ingress core interface inside MPLS tunnels. So as long as the EXP markings of packets belonging to L3 and RT are the same, the behavior applied to both is also the same. The EXP markings could also be different and still be mapped to the same queue. However, there is no gain in doing this, and in fact it would be a disadvantage, because we would be consuming two out of the eight possible different EXP markings to identify two types of traffic that require the same behavior.

As a side note, the only possible change that might be required is adjusting the transmit rate of Q4, because this new service may lead to more traffic belonging to that class of service crossing this router.

On the PE router, the set of tools used remains the same. The changes happen in whichever fields are evaluated by the QOS tools, such as the classifier. As previously discussed, traffic belonging to the RT class of service is identified by its User Priority marking. So if L3 traffic is identified, for example, by its DSCP field, the only change is to apply a different classifier according to the type of service, as illustrated in Figure 9.22.

In this scenario, the integration of the new service is straightforward, because the requirements can be mapped to a behavior aggregate that is already implemented.

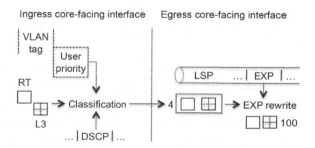

Figure 9.22 Adding a new L3VPN service

However, let us now assume that the new service has requirements for which no exact match can be found in the behavior aggregates that already exist in the network.

The obvious easy answer for such scenarios is to simply create a new behavior aggregate each time a service with new requirements needs to be introduced. However, this approach is a not scalable solution and causes various problems in the long run, such as exhaustion of EXP values and an increase in jitter because the scheduler has to jump between a greater number of queues.

There are several possible solutions. One example is the use of a behavior aggregate that has better QOS assurances. Another is to mix traffic with different but similar requirements in the same queue and then differentiate between them inside the queue by using the QOS toolkit.

Previously in this case study, we illustrated that using WRED profiles inside Q1 allows DA green and DA yellow traffic to have access to different queue lengths, even though both are mapped to the same queue.

The same type of differentiation can be made for the queue transmit rate. For example, we can limit the amount of the queue transmit rate that is available for DA yellow traffic. This can be achieved by applying a policer in the egress direction and before the queuing stage.

A key point that cannot be underestimated is the need for network planning for the services that need to be supported both currently and in the future. It is only with such analyses that synergies between different services can be found that allow the creation of the minimal possible number of behavior aggregates within the network.

9.15 Multicast Traffic

Services that use multicast traffic are common in any network these days, so let us analyze the implications of introducing multicast to the QOS deployment presented so far, by assuming that the BU class of service carries multicast traffic.

The set of QOS tools that are applied to BU traffic are effectively blind regarding whether a packet is unicast or multicast or even broadcast. The only relevant information are the packets' QOS markings and their arrival rate at the ingress PE router. However, the total amount of traffic that is present in the BU class of service affects the dimensioning of the length and transmit rate of Q2, into which the BU traffic is placed. The result is that there is a trade-off between the queuing resources required by Q2 and the effectiveness of how multicast traffic is delivered across the network.

As a side note, it is possible for classifiers to have a granularity level that allows the determination of whether a packet is unicast or multicast, so that different classification rules can be applied to each traffic type, as needed, as explained in Chapter 5.

Let us now analyze how to effectively deliver multicast traffic across the network in this case study. Within the MPLS realm, the most effective way to deliver multicast traffic is by using a point-to-multipoint (P2MP) LSP, which has one source and multiple destinations. The efficiency gain is achieved by the fact that if a packet is sent to multiple sources, only a single copy of the packet is placed in the wire. Figure 9.23 illustrates this behavior. In this scenario, a source connected to PE1 wishes to send multicast traffic toward two destinations, D1 and D2, that are connected to PE2 and PE3.

The source sends a single multicast packet X, and a single copy of that packet is placed on the wire inside the P2MP LSP. When a branching node is reached, represented by router B in Figure 9.23, a copy of packet X is sent to PE2 and another copy is sent to PE3. However, only one copy of packet X is ever present on the links between any two network routers. Achieving such efficiency in the delivery of multicast traffic translates into having no impact on the QOS deployment presented thus far. In a nutshell, the better the delivery of multicast traffic across the network is, the less impact it has on the QOS deployment.

If P2MPLSP were not used, the source would have to send two copies of packet X to PE1, which would double the bandwidth utilization. As such, the transmit rate requirements of Q2 would increase and it would need a larger queue length. As a side note, P2MP LSPs are not the only possible way to deliver multicast traffic. It was chosen for this example because of its effectiveness and simplicity.

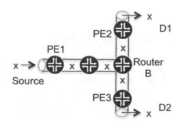

Figure 9.23   Delivery of multicast traffic using P2MP LSPs

9.16   Using Bandwidth Reservations

So far, we have considered one LSP carrying all the traffic between two PE routers. We now expand this scenario by adding bandwidth reservations and considering that the LSPs are aware of the class of service.

In this new setup, traffic belonging to each specific class of service is mapped onto a unique LSP created between the two PE routers on which the service end points are located, and these LSPs contain a bandwidth reservation, as illustrated in Figure 9.24. Also, the LSPs maintain 100% path control because they use strict ERO, as mentioned previously in this case study.

Establishing LSPs with bandwidth assurance narrows the competition for bandwidth resources, because the bandwidth value reserved in an LSP is accessible only to the class of service traffic that is mapped onto that LSP. However, this design still requires admission control to ensure that at the ingress PE router, the amount of traffic placed into that LSP does not exceed the bandwidth reservation value.

Bandwidth reservation in combination with the ERO allows traffic to follow different paths inside the network, depending on their class of service. For example, an LSP carrying RT traffic can avoid a satellite link, but one carrying BE traffic can use the satellite link.

So using bandwidth reservations offers significant gains for granularity and control. But let us now consider the drawbacks. First, there is an obvious increase in the complexity or network management and operation. More LSPs are required, and managing the link bandwidth to accommodate the reservations may become a challenge on its own.

Typically, networks links have a certain level of oversubscription, following the logic that not all customers transmit their services at full rate all at the same time. This can conflict with the need to set a strict value for the LSP's bandwidth.

Also, how to provision and plan the link bandwidth to account for possible link failure scenarios becomes more complex, because instead of monitoring

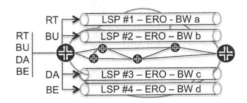

Figure 9.24   Using LSPs with bandwidth reservations and ERO

links utilization, it is necessary to plan the different LSP's priorities in terms of setup and preemption. So as with many things inside the QOS realm, the pros and cons need to be weighed up to determine whether the gains in control and granularity are worth the effort of establishing bandwidth reservations.

9.17   Conclusion

This case study has illustrated an end-to-end QOS deployment and how it interacts with other network processes such as routing. The directions we have decided to take at each step should not be seen as mandatory; indeed, they are one of many possible directions. As we have highlighted, the choice of which features to use depends on the end goal, and each feature has pros and cons that must be weighed. For example, using LSP with bandwidth reservations has a set of advantages but also some inherent drawbacks.

Further Reading

Kompella, K. and Rekhter, Y. (2007) RFC 4761, Virtual Private LAN Service (VPLS) Using BGP for Auto-Discovery and Signaling, January 2007. https://tools.ietf.org/html/rfc4761 (accessed September 8, 2015).

Lasserre, M. and Kompella, V. (2007) RFC 4762, Virtual Private LAN Service (VPLS) Using Label Distribution Protocol (LDP) Signaling, January 2007. https://tools.ietf.org/html/rfc4762 (accessed August 21, 2015).

10

Case Study QOS in the Data Center

QOS in the Data Center—is that an impossible equation? QOS is all about making the best use of the available bandwidth in relation to need and that sometimes means selectively punishing someone while protecting someone else. In brief, it's overprovisioning as a service, which most often has been applied to end users. But what is the overprovisioning in a Data Center, and what applications or traffic types can be seen as better or worse compared to others? Drops and TCP retransmissions, are those acceptable behaviors in today's Data Center? And with today's Data Centers where most of the traffic volume is actually East–West, that is, between bare-metal servers and virtual machines (VM) within the same Data Center (not the legacy North–South traffic model), as illustrated in Figure 10.1, what is an important flow versus a not so important one? One last fact to consider is that end users are hard to identify and applications inside the Data Center are more or less equally important.

10.1 The New Traffic Model for Modern Data Centers

The legacy traffic pattern for Data Centers has been the classical client–server path and model. The end user sends a request to a resource inside the Data Center and that resource computes and responds to the end user as illustrated in Figure 10.2. The Data Center is designed more for the expected North to South traffic, rather than the possible traffic that traverses between the racks. The

QOS-Enabled Networks: Tools and Foundations, Second Edition. Miguel Barreiros and Peter Lundqvist.
© 2016 John Wiley & Sons, Ltd. Published 2016 by John Wiley & Sons, Ltd.

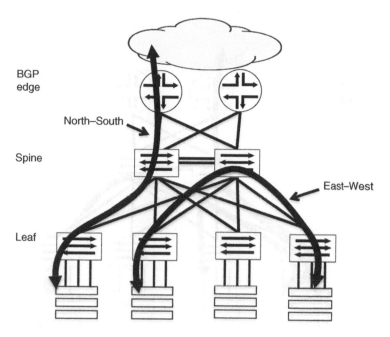

Figure 10.1   North–South versus East–West traffic

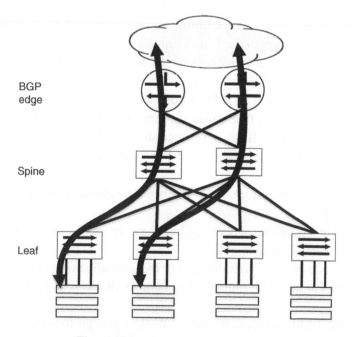

Figure 10.2   North–South traffic model

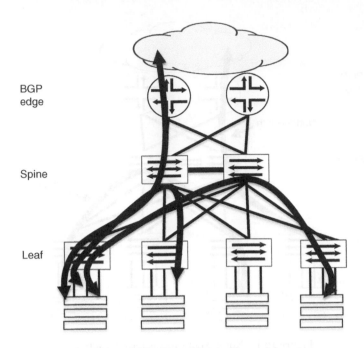

Figure 10.3   East–West traffic model

design focus on transport, not really on residential applications. This results on traffic patterns that do not always follow a predictable path due to asymmetric bandwidth between layers in the Data Center.

The evolution that has taken place is the increased machine-to-machine traffic inside the Data Center. The traffic proportion regarding North–South versus East–West is now at a ratio of at least 80% in favor of East–West compared to North–South. The result is that traffic patterns exist for only that which resides within the Data Center itself, such as those shown in Figure 10.3.

The reasons for this increased East-to-West traffic are:

- Applications are much more tiered where web, database, and storage interact with each other.
- Increased virtualizations where applications are easily moved to wherever compute resources are available.
- Increased server-to-storage traffic due to the separation of compute nodes and storage nodes demanding much higher bandwidth and scalability.

How does the tiered applications work? Well, if you go to a website with the intention of buying a book like this, then there are lots of windows that pop up on your web screen window. There is of course information about the book, but also review information, suggestions of other similar QOS book titles (none as good as this one, of course), shopping carts, location-based advertising, and so on. Your shopping session results in lots of subsessions where pictures are grabbed from one server, samples from another book, and so forth. There is also lots of synchronization traffic between servers running the same applications and data.

Another driver of the East–West traffic model is the increased usage of virtualization, which in brief increases the utilization usage per hardware in the Data Center. This means VMs and applications become more mobile since any free computing resources can be used in the Data Center. One such example is VMware vMotion that allows a VM to be moved wherever resources are available, illustrated in Figure 10.4. The result is that applications are not hosted within the same rack any longer, and instead the network has to support any-to-any communication regarding paths and predictable performance.

Another reason for the increased East–West traffic is the vast crunching of data. For example, when you surf to a website on your laptop, it's not just by happenstance that offers are suddenly presented to you that you have nothing to

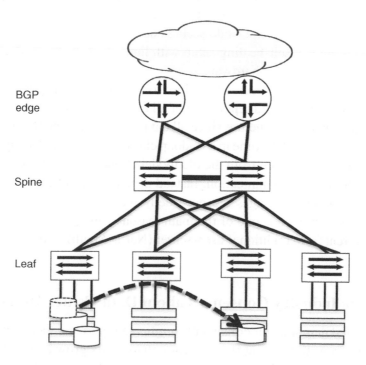

Figure 10.4   Applications run where resources are available

do with your current session. Lots of information is taken from your browser, location, profile, and even prior web-surfing history. This concept is called "big data" and is one of the most bandwidth intense driven solutions in a Data Center of today.

A software program called Hadoop is commonly used for these big data scenarios. Originally from an open-source software, the Apache Hadoop algorithm allows distributed processing of large data sets across clusters of machines using the same software library. Hadoop is designed to scale from single servers to thousands of servers where each server does the local computation and storage. What makes Hadoop scalable is its architecture. The Hadoop Distributed File System (HDFS) splits files into blocks and distributes these to other machines in the cluster called *DataNodes*. Processing is handled by the MapReduce function that distributes the processing ability to where the data resides, allowing high-redundancy, same-time parallel processing of large amounts of data. The MapReduce engine consists of a JobTracker to whom the client applications submit jobs. JobTracker then pushes jobs as tasks to the TaskTrackers that reside on each of the DataNodes in the cluster. All the DataNodes are handle by a NameNode that secures the validity of the DataNodes and thereby the actual data itself. Hadoop is designed to detect and handle failures at the application layer in any of the machines that may be prone to failure. JobTracker just assigns job to other nodes, and NameNode secures data that's been available on other nodes. The result is a self-healing setup with limited needs, for example, dual-homed links and dedicated hardware.

In big data, jobs gets processed as close to the local data as possible (rack), however the distribution and replication of all this data results in a heavy burden on the Data Center infrastructure since all data is replicated and that replication happens more or less all the time in a Hadoop cluster, as illustrated in Figure 10.5.

This is a clear difference from the old model of one interaction being limited to one North–South communication. Instead there are tons of East–West traffic involved before any messages are even delivered back to the end user. This East–West phenomenon is one of the main reasons why QOS in a Data Center is very different from traditional user–server (North–South) traffic.

10.2   The Industry Consensus about Data Center Design

The increased demands on consistent, predictable performance and latency, while at the same time being flexible enough to easily grow and scale in case of need, puts new demands on Data Center topology and structure. A well-thought

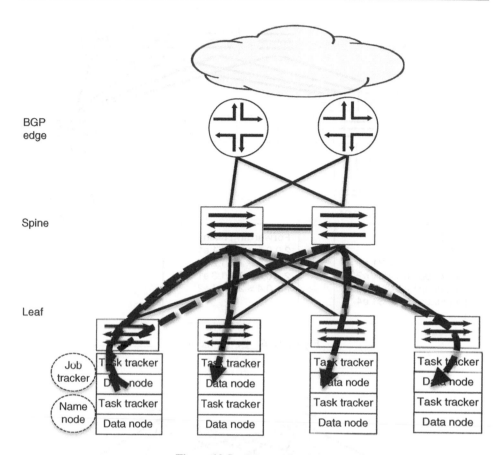

Figure 10.5   Data replication

out design is one where traffic paths are predictable and resources are well used. There is an industry consensus that a Spine-and-Leaf design, with a ECMP load balanced mesh, creates a well-built *fabric* that both scales and at the same time is efficient and most of all, predictable. That means no ring structure or asymmetric bandwidth clustering. With a two-tier Spine and Leaf design, paths are never more than two hops away resulting consistent round-trip time (RTT) and low jitter with flow/session hash, shown in Figure 10.6.

One Data Center design rule is to keep the Spine as clean as possible with no resources like servers attached to it. Instead, not just servers are allocated to the leafs, but also other resources like edge routers, load balancers, firewalls, etc., shown in Figure 10.7. By keeping this clean spine design the network is consistent and predictable regarding behavior and performance, while any failing equipment is easy to isolate and replace.

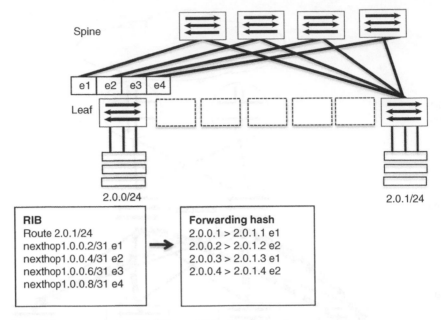

Figure 10.6   ECMP mesh

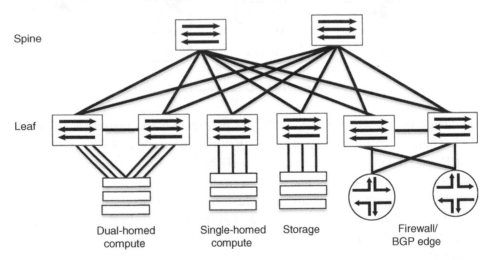

Figure 10.7   Functions and compute on the Leaf only

Regarding redundancy, the best redundancy is the one that is always in use. If you're using an active–passive design for dual-homed servers, it's a receipt for possible issues once the passive needs to take off for a faulty active. If both links are active, then not just redundancy has been verified as part of the daily performance, bandwidth is also increased. A common feature to achieve this is using Multichassis Link Aggregation (MLAG) that allows the same LAG from hosts or switches to be terminated on dual switches in a active–active design without any blocked ports or passive state. The same principle goes for the links between Spine and Leaf—if all links are active with the same metric then redundancy is part of the actual ECMP forwarding mesh. A failed link is just a temporary reduction of bandwidth and nothing else.

One trend in large Data Centers is to deploy IP as the protocol between Leaf and Spine. This limits the Layer 2 domain and possible Layer 2 protocols that are not designed for twenty-first century networking. Spanning Tree Protocol is, by its design, a blocking architecture, whatever variations of it are used, and it fits poorly in a Leaf and Spine ECMP design. Layer 2 on the Leaf only uses MLAG or cluster functionality.

10.3   What Causes Congestion in the Data Center?

Experts and designers vary on what is the best way to implement QOS functions in the Data Center. However, before discussing a cure, let's identify the symptoms. One obvious question is *What are the congestion scenarios in the Data Center?* Let's focus on the East–West traffic and identify QOS scenarios where packets are either dropped or delayed causing sessions rate being severely decreased inside the Data Center itself.

The bottleneck scenarios are:

- Oversubscribed networks with bursts greater than available bandwidth
- Multiple nodes trying to read/write to one node resulting in TCP performance challenges
- Servers sending pause frames to the network layer, which in turn, causes congestion (a phenomena, together with flow control, detailed in Chapter 4)

10.3.1   Oversubscription versus Microbursts

No Data Center is built with the assumption and acceptance of latent bottlenecks. Instead, it's when packet bursts occasionally go beyond available bandwidth. It is common practice in the Data Center to oversubscribe the trunk between the Leaf and Spine compared to the access port bandwidth—a $1:2$ ratio means that the summary of port bandwidth intended for hosts are doubled compare to ports primary designed for trunks. A common oversubscription is in

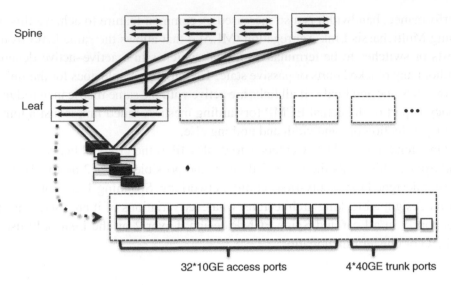

Spine

Leaf

32*10GE access ports   4*40GE trunk ports

Figure 10.8   Leaf switch oversubscription by design

the ratio of $1:2$ or $1:3$. Of course there are variations here with more or less oversubscription. The following shows an illustration of a Leaf switch with 32*10GE access ports to hosts and 4*40GE trunk ports to connect to spine. This means a $2:1$ oversubscription rate to hardware architecture for traffic leaving the trunk ports, as illustrated in Figure 10.8.

To avoid too much oversubscription by hardware design is of course simply not to oversubscribe. If the design is not to oversubscribe and instead use a $1:1$ ratio regarding bandwidth, and if using the hardware earlier described, then do not use more access ports than the switch's weak link, which is the trunk speed. If the available trunk bandwidth is 160 Gbps, then maximum access bandwidth can only be 160 Gbps, that is, one half of the access ports.

Another oversubscription scenario is the scenario called *microbursts*. The phenomenon has been known for a long time in the mobile networking world but is a relatively recent reality in the IP and Data Center arena. A microburst event is when packet drops occur despite there not being a sustained or noticeable congestion upon a link or device. The causes for these microbursts tend to be speed mismatch (10GE to 1GE and 40GE to 10GE). The more extreme the speed mismatch, the more dramatic the symptom. Note it's not congestion due to fan-in or asymmetric speeds oversubscription, but it's the simple fact that higher bandwidth interfaces have higher bit rates than lower bandwidth interfaces. A typical example in the Data Center world is with asymmetrical speed upgrades. The hosts are on the 1GE access links, but the storage has been upgraded

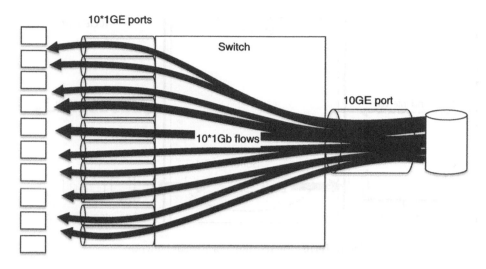

Figure 10.9   Speed conversion

to 10GE link. In theory, this appears not to be a problem as seen in the next illustration Figure 10.9.

However, the real challenge here is the bit rate difference between 1GE and 10GE. In brief, ten times (10*) the difference results a short burst of packets toward the 1GE interface. In case there are small interface buffers, or lots of flows that eat up the shared pool of buffering, there will be occasional drops that are not easy to catch if you are not polling interface statistics frequently enough. There is only one way to solve this and that is to be able to buffer these bursts as shown in the next illustration Figure 10.10.

Another parameter to be aware of is the serialization speed difference between 1GE versus 10GE and 10GE versus 40GE. For example, the time it takes to build a 100-byte packet for 1GE interface is 800 ns, and the same 100-byte packet for 40GE takes 20 ns, which can constrains the buffering situation in a topology even more with speed mismatch shown in Figure 10.11.

One way to limit microbursts is to have the same bit rate and thereby serialization end to end within the Data Center. That is, try to avoid too much difference in the interfaces speed. For example, in a Hadoop cluster it's suboptimal to run servers with different bandwidth capabilities since the concept is to have all the data replicated to several NameNodes. Then, if the nodes have different interface speeds, obviously microburst situations can easily occur. A common speed topology can be achieved by using multispeed ports (most often referred as MTP). In brief, it's a single cable consisting of 12 fibers resulting in a 40-Gbps port can be 4*10 Gbps channel interfaces. Thereby the same bit rate can be achieved as shown in the following illustration (Figure 10.12).

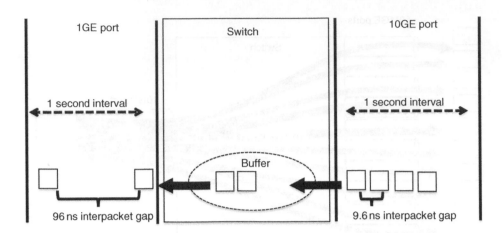

Figure 10.10   Bit rate difference needs buffering

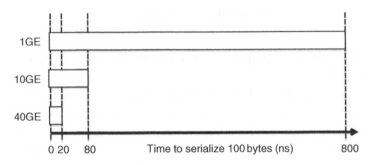

Figure 10.11   Serialization difference interface speed 100 byte

Still, the only really efficient and same-time solution that can flexibly handle traffic bursts inside the Data Center is to have buffering capabilities in both the Spine and the Leaf layer.

10.3.2   TCP Incast Problem

Some might consider TCP Incast as a fancy term for congestion due to multiple nodes trying to read/write to the same node. While it is a many-to-one scenario, it's more than that since it affects the TCP protocol. In brief, what happens is:

1. Many nodes access the same node in a tiered application scenario or over utilized DataNode in a Hadoop cluster.
2. Packets will be dropped or delayed upon buffer exhaustion.
3. Dropped packets or packets RTT exceeding the RTO will cause retransmission.

Retransmission is something you don't want for East–West traffic for all the obvious reasons but also because the higher utilization leads to unfairness across

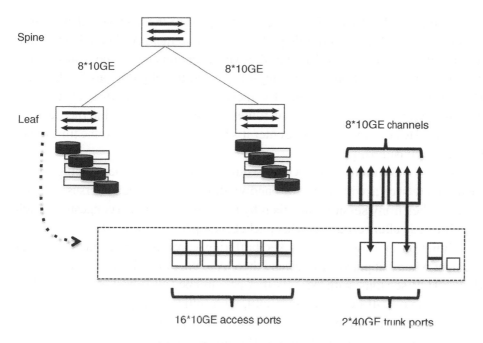

Spine

8*10GE   8*10GE

Leaf

8*10GE channels

16*10GE access ports   2*40GE trunk ports

Figure 10.12   Bit rate consistency

flows. And this is a vicious cycle where throughput suffers due to TCP congestion behavior. The result can be that the network's overall statistics look under-
utilized regarding statistics from network equipment, since those counters only
see the traffic that actually can be delivered and do not show the TCP backoff
symptoms (the details regarding TCP backoff mechanisms have been described
in depth earlier in Chapter 4).

Let's briefly review TCP's congestion behavior. The sender's TCP congestion window (CWND) described earlier in this book relies upon the
receiving TCP acknowledge (ACK) packets in a timely manner from the
receiver to be able to adjust the speed of transmission of packets. The rate
is simply the number of packets transmitted before expecting to receive
these TCP ACKs. When an accepted packet has been lost, the retransmission timeout (RTO) settings on the servers determine how long TCP waits
for receiving acknowledgment (ACK) of the transmitted segment. In a
nutshell, it's the expected round-trip time (RTT). If the acknowledgment
isn't received within this time, it is deemed lost. This is basically the TCP
rate control and congestion management applied on top of the available
network bandwidth. If there is insufficient bandwidth in the network to
handle these bursts, packets are dropped in order to signal the sender to
reduce its rate.

So what now? Why not just increase the speed to those servers that have to feed others, or at least secure they are on the same rack. And here is the catch described earlier, in the modern world of virtualization and Hadoop clusters, it's not that trivial to find out these many-to-one and thereby obvious bottlenecks. However, there is a known cure for these bursts—apply buffers on the networking equipment—a lesson that routers in the core of the Internet learned years ago. However, in the Data Center it's been a bad habit for some time to not have enough buffering capacity.

But why in the name of Bill Gates would small buffers on networking equipment create such a wide range of bandwidth resulting in lucky versus unlucky flows? When many sessions, and thereby flows, pass through a congested switch with limited buffer resources, what packet is dropped and what flow is impacted is a function of how much packet buffering capacity was available at that moment when that packet arrived and thereby a function of lucky versus unlucky flows. TCP sessions with packets that are dropped will backoff and get less share of the overall network bandwidth, with some unlucky flows getting their packets dropped much more than others resulting in senders congestion window half, or even end up, in a TCP slow start state. Meantime, the lucky sessions that just by luck have packets arriving when packet buffer space is available do not drop packets and instead of slowing down will be able to increase their congestion window and rate of packets as they grab unfairly at the amount of bandwidth. This behavior is most often referred as "TCP/IP Bandwidth Capture Effect," meaning that in a congested network with limited buffer resources, some session's flows will capture more bandwidth than other flows, as illustrated in Figure 10.13.

So how to avoid this behavior with lucky versus unlucky flows? Well, one traditional way is to implement RED schemes described earlier but ultimately that will also result in retransmissions and the TCP congestion management vicious cycle. RED originally was designed for low speed connections and is also more suitable to end users than East–West traffic server connections. Another solution discussed is to use UDP inside the Data Center, but the result is that the delivery assurance moved from the TCP stack to the application itself. Some advocate for a new form of TCP, Data Center TCP (DTCP) that uses the Explicit Congestion Notification (ECN) bits field, and by marking, detects congestion and speeds up the feedback process in order to reduce the transmission rate before too many drops and the TCP mechanics kicks in. In brief, a Layer 3 way of Layer 2 flow control. However, the true story is that the best way to handle TCP Incast scenarios is with buffering capabilities on the networking equipment, the same conclusion as with microbursts.

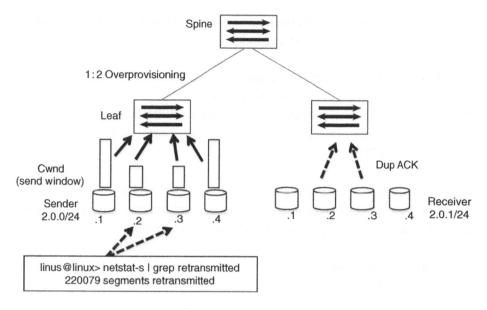

Figure 10.13   TCP Incast

10.4   Conclusions

There is an ongoing discussion that too much buffer is just as bad as drop packets. The discussion is that buffers should be small: *You can use larger buffers to mask the issue, but that just increases latency and causes jitter. In extreme cases, you can end up with the situation of packets arriving seconds latter. Dropping packets is not the end of the world, but it puts a limit on how large the latency and jitter can grow.*

There is absolutely some validness in this italicized statement regarding end user's quality experience for applications like VoIP and others. But for East–West server-to-server traffic, there is another story. If you lack buffers and cannot absorb burst, you end up in retransmissions, and that is really a consequence of inefficient management of resources and TCP performance. The default RTO on most UNIX/Linux implementations residing in Data Centers is around 200 ms. Tuning RTO is outside the scope of this book and it is not a trivial task, but let's discuss what is the ultimate buffer size for networking equipment inside a Data Center if the majority of application and traffic is TCP based, since congestion results in TCP backoff and smaller TCP CWND. The perfect scenario is if the switch can buffer an entire TCP window for all possible congested session flows, then no retransmissions are needed!

Let's play with numbers and plug this equation: the number of flows without drops = Total packet memory/TCP window:

- The Max TCP window per flow equal to 64 KB
- Servers sustain max 100 simultaneous flows
- Packet memory needed per port = 64 KB × 100 = 6400 KB
- Total packet memory needed (memory per port × number of ports) = 6400 KB × 32 = 256 MB

So if buffers is the name of the game for optimal performance for Data Center East–West traffic, let's take a look at the most common hardware architectures for them on networking equipment inside the Data Center:

- Dedicated per port buffer
- Shared-pool buffer
- Virtual output queuing (VOQ)

The architecture with a dedicated buffer allocated per port is a simple construct regarding its architecture. In brief, you quote the memory banks to each port and thereby each port has its own resource. It's a good design if all the ports are expected to talk to each other simultaneously, thereby limiting possible greedy ports that might affect the performance of others. Its drawback is that it drains lots of memory and there's also some possible backpressure scenarios that are hard to solve since that would demand multiple ingress and egress physical buffers. In a case where memory is well allocated, it's mostly in an off-chip design scenario resulting in possible extra delay and costly extra components like extra processors to handle the queuing memory.

The shared-pool buffer architecture assumes that congestion occurs sporadically only to a few egress ports and, realistically, never happens on all ports simultaneously. This allows a more centralized chip buffer architecture that is a less costly design compared to dedicated memory. However, the shared-pool buffer architecture demands a complex algorithm so that the buffer is dynamically shareable and weighted toward congested ports when backpressure from these ports occurs due to congestion. No port and queue is allowed holding on the buffers for too long or too much just in case of peaks and burst. It also demands tuning thresholds for ingress versus egress buffering and port versus shared-pool volume.

The VOQ is a technique where traffic is separated into queues on ingress for each possible egress port and queue. It addresses a problem discussed in the earlier chapters with head-of-line blocking (HOL). In a VOQ design each input port maintains a separate queue for each output port. VOQ does demand some advanced forwarding logics like cell-based crossbars and advanced scheduling algorithms. And the VOQ mechanism provides much more deterministic throughput at a much higher rate than crossbar switches running without it, but VOQ demands much more advanced scheduling and queue algorithms, and the hardware designs can be costly. But like most things in life, there are no simple solutions and there are no such things as a free lunch, even at the buffer buffet.

Further Reading

Bechtolsheim, A., Dale, L., Holbrook, H. and Li, A. (2015) Why Big Data Needs Big Buffer Switches, February 2015. https://www.arista.com/assets/data/pdf/Whitepapers/BigDataBigBuffers-WP.pdf (accessed September 8, 2015).

Chen, Y., Griffith, R., Liu, J., Katz, R.H. and Joseph, A. (2009) Understanding TCP Incast Throughput Collapse in Data Center Networks, August 2009. http://yanpeichen.com/professional/TCPIncastWREN2009.pdf (accessed August 19, 2015).

Das, S. and Sankar, R. (2012) Broadcom Smart-Buffer Technology in Data Center Switches for Cost-Effective Performance Scaling of Cloud Applications, April 2012. https://www.broad com.com/collateral/etp/SBT-ETP100.pdf (accessed August 19, 2015).

Dukkipati, N., Mathis, M., Cheng, Y. and Ghobadi, M. (2011) Proportional Rate Reduction for TCP, November 2011. http://dl.acm.org/citation.cfm?id=2068832 (accessed August 19, 2015).

Hedlund, B. (2011) Understanding Hadoop Clusters and the Network, September 2011. http://bradhedlund.com/2011/09/10/understanding-hadoop-clusters-and-the-network/ (accessed August 19, 2015).

Wu, H., Feng, Z., Guo, C. and Zhang, Y. (2010) ICTCP: Incast Congestion Control for TCP in Data Center Networks, November 2010. http://research.microsoft.com/pubs/141115/ictcp.pdf (accessed August 19, 2015).

11

Case Study IP RAN and Mobile Backhaul QOS

Radio access networks (RANs) connect mobile base stations to the mobile backhaul network. RANs have evolved from second-generation (2G) networks with GSM handsets to third-generation (3G) networks, which introduce IP. However, 3G networks do not offer true IP-based service. Rather, SSGN tunnels the data portion of the traffic to general packet radio service (GPRS) routers, which act as gateways to IP-based networks. The next phase, fourth-generation (4G) networks, commonly called Long-Term Evolution (LTE), introduces more IP into mobile backhaul networks, transforming RANs into IP RANs. In LTE networks, voice packets are encapsulated into IP packets and are transmitted over IP RAN, not over the legacy public switched telephone network (PSTN) as is the case with 3G networks.

This case study examines the recently evolved LTE network, with a focus on packet-based QOS. It starts by discussing the components of 2G and 3G networks, how traffic is carried on these networks, and how they have evolved to LTE. The case study then describes the LTE network components and traffic and offers guidelines and suggestions for using QOS.

11.1 Evolution from 2G to 4G

This book focuses on QOS in packet-based networks, and this chapter presents a QOS case study for IP-based RAN as part of a mobile backhaul network. However, a brief introduction of 3G network evolution to 4G and LTE is

QOS-Enabled Networks: Tools and Foundations, Second Edition. Miguel Barreiros and Peter Lundqvist.
© 2016 John Wiley & Sons, Ltd. Published 2016 by John Wiley & Sons, Ltd.

necessary to describe the fundamental changes that occur with LTE and the packet transport for this service. We must discuss some basic mobile functions and features before explaining the QOS scenario for IP RAN and mobile backhaul. This discussion touches on the packet-handling services in 3G and 4G networks. Global System for Mobile Communications (GSM), Code Division Multiple Access (CDMA), and any other pure mobile voice and signaling architectures are described only briefly. Also, the 3rd Generation Partnership Project (3GPP) evolution is covered only in generic terms.

11.2   2G Network Components

2G mobile networks and GSM are synonymous, but GSM is widely seen as one of the key components within 2G mobile phone systems. With GSM, both signaling and speech channels are digital. This facilitates the widespread implementation of data communication applications into the mobile network. GSM is a cellular network, which means that mobile phones connect to it by searching for cells in the immediate area. A cell is a part of a base station, and several base stations form the cellular network. GSM networks operate in a number of different carrier frequency bands, normally within the 900- or 1800-MHz band, although this varies from country to country. The frequencies are divided into timeslots. The elements in a GSM mobile network can be seen in Figure 11.1.

The mobile system (MS) includes a mobile equipment (ME) and a subscriber identity module (SIM). The SIM contains the subscriber's International Mobile Subscriber Identity (IMSI), which carries information about the area code, the country code, the identity, and so forth. The SIM also holds the subscriber's cryptographic key, which is used for authentication and security.

The Base Transceiver Station (BTS), which is basically a radio receiver, handles functions related to radio access between the mobile phone and the

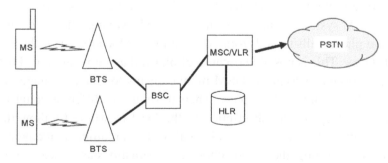

Figure 11.1   Key elements of a GSM network

BTS. Each BTS covers a cell, which is a radio access area. Normally, BTSs are grouped in an overlapping design so that they span several cell areas. A group of BTSs is commonly referred to as a "site." In IP terminology, the BTS is the GSM's customer premises equipment (CPE) gear.

The Base Station Controller (BSC) offloads the mobile switching center (MSC) of radio-related items, such as radio call setup, administration of channels, and handover (roaming) between cells (BTS coverage).

The MSC is the actual traffic or call route switch and performs the switching between the 64-kbps channels. The MSC is also responsible for monitoring and registering calls.

The Visitor Location Register (VLR) is a database located in the MSC. It maintains a list of the subscribers for those mobile phones that currently exist within the MSC coverage area.

The Home Location Register (HLR) is the main database that maintains a list of each GSM operator's subscribers. The HLR performs several functions related to maintaining and administering mobile services and is supported by other databases, including:

- Authentication Center (AUC): All GSM operators maintain this database, which contains information for subscriber authentication, including all subscribers' cryptographic keys.
- Equipment Identity Register (EIR): When subscribers log in to a GSM operator, the operator registers the IMSI that exists on the MS. The operator also transmits the unique mobile phone identifier, called the International Mobile Equipment Identity (IMEI), to enable an operator to lock out "missing" units.
- Central Equipment Identity Register (CEIR): In this database, operators report all IMEIs that should be locked out from the network.

GSM is a cellular telephony system that supports mobility over a large area. Unlike cordless telephony systems, it provides roaming and handover functionality. The GSM system is capable of international roaming; that is, it can make and receive phone calls to and from other nations or locations as if the user had never left home. This is possible because of bilateral agreements signed between the different operators to allow GSM mobile clients to take advantage of GSM services with the same subscription when traveling to different countries and locations, as if they had a subscription to the local network. To allow this, the SIM card contains a list of the networks with which a roaming agreement exists. When a user is roaming, the mobile phone automatically starts a search for a network stipulated on the SIM card list. The choice of a network either is performed automatically or can be selected by the user. The home Public LAN Mobile Network (PLMN) is the network to which the user is subscribed, while the visited PLMN is the one in which the user is roaming. When the user is moving from one cell to the other during a call, the radio link between BTS and the MS can be replaced by another link to another BTS. The continuity of the call can be performed in a seamless way for the user. In brief, this is the handover process.

GSMs have been improving year by year, with new features and enhancements to existing features defined in annual releases named after the year of introduction (release 96, 97, 98, 99, and so forth). The GSM responsibilities have been managed by the 3GPP. An example of an enhancement managed by 3GPP is GPRS; CDMA is different from GSM. CDMA is not discussed in this chapter because there is no value when describing the QOS aspects of 4G networks in pointing out the differences and similarities between GSM and CDMA.

11.3 Traffic on 2G Networks

The traffic on 2G networks passes through timeslot- or cell-based networks as it heads towards the PSTN. The RAN provides no packet processing for any application. All traffic is handled within 64-kbps timeslots or SDH/ATM trunks; only circuit-switched mode is used for any type of traffic. Figure 11.2 shows the 2G network transport elements.

11.4 3G Network Components

The 3G network represents an evolution of the 2G network and GSM. 3G networks allow data transmission in packet mode rather than the circuit- and timeslot-switched modes of 2G networks. Packet mode is made possible by

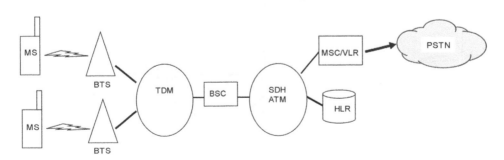

Figure 11.2 Transport elements of the 2G network

GPRS. GPRS still uses the GSM principles of a radio interface and its notions of timeslots. The system design goal of GPRS is to provide higher data-rate throughput and packet-switched services for data traffic without affecting pure voice and speech traffic and minimizing the impact on the existing GSM standard and networks.

The GPRS network architecture reuses the GSM network nodes such as MSC/VLR, HLR, and BSC. It introduces two new network nodes for the transport of packet data: the gateway GPRS support node (GGSN) and the serving GPRS support node (SGSN). GPRS also defines new interfaces between the GSM network nodes and the different elements of the GPRS core network (see Figure 11.3).

This book makes no distinction between acronyms that mean more and less the same thing. An example is 3G versus Universal Mobile Telecommunications System (UMTS). UMTS is one of the 3G mobile telecommunications technologies, which is also being developed into a 4G technology. With UMTS, the BTS or NodeB still handles the radio access. Both packet-oriented services such as GPRS and circuit-switched services such as voice can use the same cell from the NodeB.

The BSC or Radio Network Controller (RNC), as it is called in 3G, has the same function as in 2G, although some 3G functionality has been added to route voice traffic to circuit-switched networks and data traffic to packet-based networks. RNC uses acronyms for its interfaces towards NodeB, all of which start with "lu" and a letter, for example, "b." Figure 11.4 shows the interface naming for radio elements.

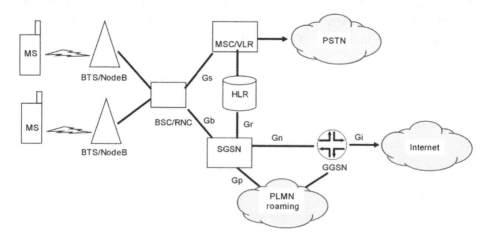

Figure 11.3   3G GPRS architecture

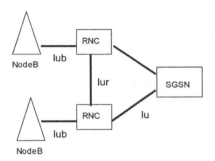

Figure 11.4   3G radio interface

The SGSN is the node that serves the MS. It delivers packets to and from the MS and communicates with the HLR to register, to authenticate, and to obtain the GPRS profile of the subscriber. The SGSN also performs accounting. The SGSN can be connected to one or several BSCs or RNCs. Also, the SGSN is the device that manages mobility if the subscriber moves, finding the user's "home" GGSN with assistance from the HLR. The SGSN functions very much the same way as the MSC does for voice. In a nutshell, the IP part from the SGSN starts with the packet service that the GPRS provides. In GPRS terms, the network interface between the SSGN and NodeB is called Gb.

A GGSN provides interworking with external Packet Data Networks (PDNs). The GGSN can be connected to one or several data networks, for example, to an operator's Internet or to a company virtual private network (VPN). It connects to the SSGN using an IP-based GPRS backbone network, commonly called the mobile backhaul. The GGSN is a router that forwards incoming packets from the external PDN to the SGSN of the addressed MS, or vice versa. Once a packet leaves the GGSN towards a PDN, it is a pure IP packet. Thus, the other routers in the Internet or company VPN see the GGSN as just another router. The GGSN communicates with other services such as RADIUS to achieve the correct user profiles, and it also performs accounting.

So how is the data traffic handled between the SSGN and GGSN? This is where the GPRS Tunneling Protocol (GTP) comes into the picture. The SSGN uses GTP to create a tunnel to the GGSN, which means that native IP processing of end user starts at the GGSN. Figure 11.5 illustrates GTP tunneling in the GPRS network.

There are two types of GTP tunnels. The tunnel for control traffic, called GTP-C, uses UDP port 2123. The second type of tunnel, for user traffic, called GTP-U, uses UDP port 2152. Figure 11.6 shows the GTP header.

Control traffic over the GTP-U is exchanged between the MS and the GPRS network components to establish a Packet Data Protocol (PDP) context, defined in RFC 3314 [1]. The traffic over the GTP-U, which is the actual user traffic, is

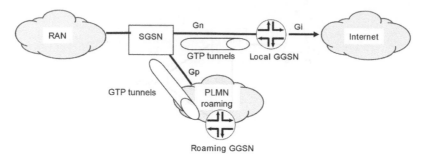

Figure 11.5 GTP tunneling

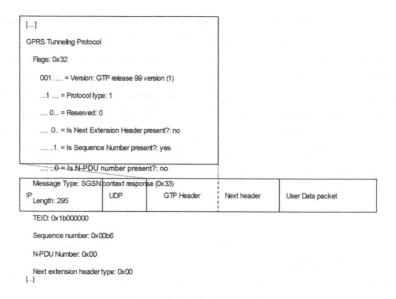

Figure 11.6 The GTP header

exchanged between the MS and the GGSN. The PDP context is the microflow entity for the user session. The mobile world does not talk about sessions; it talks instead about PDP context. Several PDP contexts can be active between the MS and the GGSN. The primary PDP context defines the connection to the GGSN, and the secondary context can have different attributes such as QOS because it can support different services. The PDP context can be seen as a mix between service and application, because each context established on the MS is assigned its own IP address. In this way, the PDP context is not similar to TCP, for example, which uses sockets and port numbers on the host itself to separate established applications from each other. Three basic actions can be performed on a PDP context: activation, modification, and deactivation. Figure 11.7 illustrates the PDP context activation.

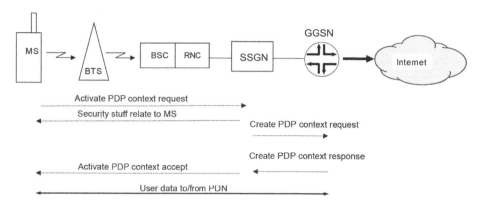

Figure 11.7   PDP context activation flow

Mobile users and their PDP contexts are grouped on the GGSN into an Access Point Name (APN). An APN is the mobile realm's parallel of a VPN. The APN manages the IP address ranges that have been allocated to each PDP context. An example of an APN is Internet access or a private VPN, towards which the GGSN tunnels packets using IPsec or GRE tunnels.

11.5   Traffic on 3G Networks

Voice traffic on a 3G network continues to be circuit switched, so in this sense it is identical to voice traffic on a 2G network. The difference lies in the data service packets, which are routed in a packet-based network structure. As we have discussed, an end user packet is not processed as a native IP packet until it reaches the GGSN. IP traffic between the SSGN and the GGSN travels in a GTP tunnel over the Gn interface or, in a roaming situation, over the Gp interface so the user can be routed over another operator's network to their home GGSN.

From a QOS perspective, this mobile traffic is clearly handled as a best-effort (BE) service, because it is transported through a tunnel and because traffic in a GPRS network is generally placed in a BE class of service. Examples of GPRS network traffic include UDP-based name queries and lookup and TCP-based HTTP downloads. Another common traffic type is webcasting traffic, which gives the impression of being a real-time service when, in fact, it is just a streamed file that is being downloaded. QOS could be used here to map the radio QOS classes to the GTP header and appropriate DSCP class in the IP header for the GTP packet, but in 3G and GPRS network, this is more theoretical rather than an actual implemented service. Figure 11.8 shows a GTP packet and delineates the QOS extension header from the DSCP part of the IP packet.

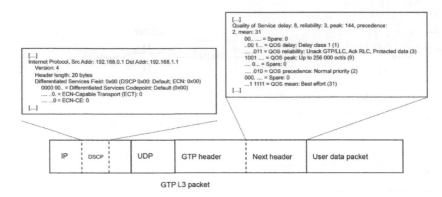

Figure 11.8   GTP QOS vs. IP QOS

GPRS traffic flows in a hub-and-spoke pattern because each subscriber is assigned to a home GGSN. This "all roads lead to Rome" traffic pattern means that the subscriber always enters and exits the network through the same GGSN regardless of whether the user is at home or is traveling.

The SSGN connected to the RNC performs a lookup of the subscriber by querying the HLR. If the subscriber is on their home GGSN, the SSGN routes the traffic to its destination. If the subscriber is roaming, the SSGN routes the traffic to their home GGSN. The result is the creation of GTP tunnels on the Gn or Gp interface in a GPRS network. Clearly, it is as hard to manage QOS functions across operators in roaming situations as it is to maintain QOS in Internet for packets that traverse service provider boundaries.

In summary, it is possible to implement IP QOS solutions in GPRS networks, but in most cases, no transparent QOS mappings between radio and IP traffic are implemented for GPRS traffic passing over the mobile backhaul on the Gn or Gp interface between the SSGN and the GGSN. The most common approach is to implement a QOS mapping on the GGSN when the packets leave the mobile network, creating a mapping between the 3GPP and other IP QOS classes such as DiffServ and IP DSCP.

11.6   LTE Network Components

3GPP, or LTE, is the next evolutionary step in radio networks. (Note that LTE and 4G are synonyms.) With LTE, the radio interface is not redirected to a PSTN, but is instead IP based. Voice is seen like any other application in an LTE network.

As an aside, LTE networks can be data-only networks that mix 3G and LTE traffic. On these networks, only the data is handled as LTE traffic, while the voice traffic remains as GSM/CDMA. Also, there are a number of variations of the voice traffic. For example, the 3GPP voice solution is generally IP Multimedia System (IMS). Discussing all the possibilities is beyond the scope of this book.

Compared to a GPRS network, the LTE network is flat. LTE has no radio control node like an RNC that collapses and aggregates the traffic. The LTE NodeB, sometimes also called the eNodeB, connects directly to an IP network, towards the Serving Gateway (S-GW), Mobility Management Entity (MME), and Packet Data Network Gateway (PDN-GW). The MME handles signaling, while the S-GW and PDN-GW handle the actual termination of tunnels and forwarding of traffic (we discuss these later).

The S-GW is the anchor point for the termination of mobile users in the packet network. It is the end point for the GTP tunnel between eNodeB and manages roaming to other networks.

The PDN-GW, or just P-GW, is the gateway between the LTE network and other networks. It acts very much like GGSN with regard to managing how it forwards users to either the Internet or a private VPN. The PDN-GW also manages the IP addressing for the end users. Note that the S-GW and PDN-GW functions can be on the same node, which often makes more sense than splitting them. Other network functions can also be running on the PDN-GW, such as NAT to map IPv6 addresses to IPv4 addresses.

The MME provides the control plane function for LTE. It is responsible for authentication and management.

System Architecture Evolution (SAE), as the LTE core network architecture is called, is an all-IP network that includes both mobile backhaul and RAN. The main components of the SAE architecture are the Evolved Packet Core (EPC), MME, S-GW, and PDN-GW.

The LTE interface names change, of course, from GPRS. Figure 11.9 illustrates the LTE S1 and Si interfaces, which are equivalent to the GPRS Gn and Gi interfaces, respectively. Figure 11.9 also separates the S-GW and PDN-GW functions to explain the interface naming simply. However, as stated earlier, the S-GW and PDN-GW can be on the same device.

One major difference between LTE and 3G, apart from the network topology, is that LTE is all IP. So, for instance, the voice is handled as VoIP. This means that all LTE operators run an IP transport network, starting from the eNodeB. The GTP tunnel between the eNodeB and the S-GW is more for administrative purposes, such as tracking and controlling users. Also, there is only one tunnel between the eNodeB and the S-GW, unlike the several tunnels present with GPRS. The TEID information differentiates the users in the GTP tunnel.

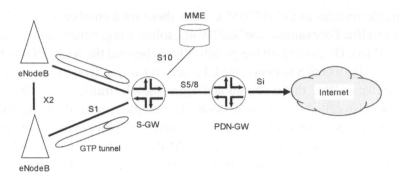

Figure 11.9   LTE architecture

The S1 and x2 interfaces shown in Figure 11.9 are considered to be RAN interfaces. These interfaces are transported over the IP RAN as part of the mobile backhaul network, possibly using public networks, such as E-Line or E-LAN. These networks are not considered secure enough to transport unencrypted data. The security is provided by IPsec, whose use is mandatory to protect the integrity for the traffic.

IPsec tunnels carry encapsulated voice and data traffic, plus signaling, between the eNodeB and the MME. The requirement for IPsec tunnel and key management is defined in the 3GPP documents TS 33.210 [2] and TS 33.31, which discuss the Layer 3 security and authentication framework. These documents require that IPsec ESP conform to RFC 4303 [3] to support integrity and replay protection and that certificate authentication be done by IKEv2.

The 3GPP mandates that IPsec be implemented on the eNodeB, and the two specifications mentioned above detail the IPsec implementation requirements. IPsec must be supported, but deploying it is not mandatory. After performing a risk assessment, the operator must decide whether to use IPsec.

However, the IPsec requirements are currently the subject of much debate. One current question is whether IPsec is necessary if the operator manages the network, end to end, from the eNodeB to the S-GW. Another question is whether IPsec is mandatory just because no authentication or encryption exists, by default, for voice traffic on an LTE network.

The extensive usage of IPsec makes traffic paths and QOS a tough challenge. In the case of a centralized S-GW using IPsec, x2 to x2 traffic termination might result in a slow or difficult handover for roaming mobile users.

While the design of LTE networks is beyond the scope of this book, it needs to be mentioned that LTE networks have a significant overhead because of the security information (based on the IPsec header) and because of the control and user tracking (based on the GTP header) of traffic between the eNodeB and

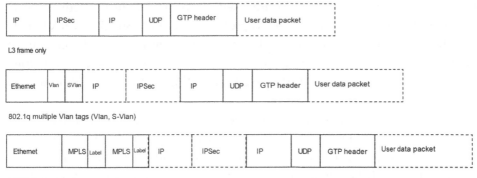

Figure 11.10   Encapsulations of LTE packets sent between NodeB and S-GW

the S-GW. To scale and segment, the network probably needs to be provisioned similarly to large access and distribution metro networks, with extensive usage of virtual LAN (VLAN) and Multiprotocol Label Switching (MPLS) stack labels to scale the provisioning of the eNodeB. Figure 11.10 shows the header overhead involved. The first packet is Layer 3 only, the second is for Ethernet aggregation using VLAN/S-VLAN, and the third shows the MPLS dual stack that is used with MPLS VPN technologies, such as L3VPN, VPLS, and pseudowires.

11.7   LTE Traffic Types

LTE networks have more traffic types than 3G networks, because all traffic is transported natively as IP packets. GTP tunneling exists, but all services and thus traffic run over the IP network. LTE traffic types can be divided into two groups: control plane and signaling, and user traffic.

On LTE networks, a vast amount of signaling and control plane traffic passes through the IP network compared with the packet-based network portion of 3G networks. Control plane and signaling traffic can be direct, indirect, or something in between.

An example of a direct control plane and signaling protocol is the mobile signaling Stream Control Transmission Protocol (SCTP), which is used to encapsulate signaling between eNodeB and MME across the Non-Access Stratum (NAS).

Indirect control plane and signaling protocols are related to the devices in the packet-based transport layer. They include the Layer 2 protocols such as ARP, spanning tree, and other resiliency protocols. Other examples are the

Ethernet OAM protocols, which are used when the dominant media type is Ethernet. Examples of Ethernet OAM protocols are IEEE 802.3ah (Link OAM) and IEEE 802.1ag (Connectivity Fault Management). The indirect control plane and signaling protocols also include the traditional Layer 3 protocols, such as the IGPs (ISIS and OSPF), BGP, and MPLS, and link-local protocols, such as VRRP and PIM.

One example of both direct and indirect control plane and signaling traffic is time signaling. In 3G networks, time synchronization is built into the transport media, for example, TDM serial links or ATM networks. Unlike TDM and cell-based networks, Ethernet carries no synchronization information in its media. When the IP RAN and mobile backhaul network is upgraded to Ethernet, the base stations are isolated from the synchronization information that used to be carried over the TDM. With LTE, voice packets in need of time and clock synchronization pass in the very same IP network in the IP RAN and mobile backhaul as data traffic, for example, HTTP traffic. The classic time service synchronization is provided by either the NTP protocol, IEEE 1588, or Synchronous Ethernet as part of power over Ethernet (IEEE 802.3af).

LTE user traffic consists of both voice and Internet applications running over TCP and UDP. Traffic is both unicast and multicast, the latter including video-conferencing and online streaming media. One point to mention is that voice is just user data-plane traffic for the LTE network. An example is SIP. A VoIP packet that uses SIP for call setup is still data, just as when the call is established with UDP packets with an RTP header.

11.8   LTE Traffic Classes

A GRPS session is referred to as a PDP context. In LTE, the equivalent is generally called the bearer, which reflects a context or session. LTE networks can have several bearers, which are used at different stages along a path in the LTE network. Figure 11.11 shows the relation between the bearer services.

The LTE network is divided into different areas, with new names, including Evolved UMTS Terrestrial Radio Access Network (E-UTRAN), which reflects the radio-only part, and EPC, which handles the packet-based networks.

A session that starts from the User Equipment (UE) and traverses the RAN and mobile backhaul towards S-GW/PDN-GW forms an EPC bearer. The focus of this discussion is on the EPC bearer, not on the end-to-end bearer, because the external barrier that is part of the end-to-end bearer might go over the Internet, crossing AS boundaries, so it needs to interact with other QOS policy domains.

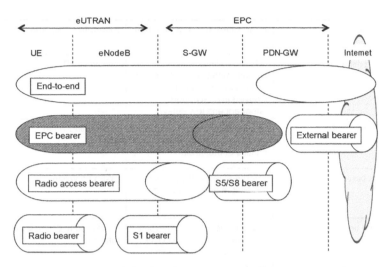

Figure 11.11   Bearer services

The assumption is that the EPC bearer runs within the same operator entity and thus within a single managed network.

Before trying to map the 3GPP and IETF QOS functions and classes for the EPC bearer, let us look at the specifics of 3GPP QOS. We briefly examine traffic separation, the types and numbers of traffic classes needed, and the QOS classes of service needed for radio traffic.

The 3GPP has defined QOS aspects in several specifications. One of the original documents, TS 23.107, "QOS Concept and Architecture," describes a framework for QOS within UMTS. This specification defines the four traffic classes shown in Figure 11.12.

The 3GPP TS 23.203 [4] and TS 23.207 [5] specifications go further, introducing QOS Class Identifier (QCI) and nine classes for LTE. QCI is an identifier that controls the QOS details on the eNodeB, and it divides the bearers into two groups: guaranteed bit rate (GBR) bearers and non-GBR bearers. Figure 11.13 describes the nine QCI groups. Figure 11.13 shows that some of the services are backward-compatible between the UMTS and LTE specifications.

Note that the 3GPP QOS models can be adapted by the IntServ and DiffServ model. This book focuses on the DiffServ and PHP QOS models; the IntServ QOS model is beyond the scope of this book.

For packet network designers, the ratio of over-provisioning is a key topic of interest, but for radio network designers, the key topic is quality of service. IP RAN and QOS designs need to satisfy both world views simultaneously. The micro versus macro angle is very much in the spotlight here regarding how to

Traffic class	Fundamental characteristics	Example of the application
Conversational real-time	Preserve time relation (variation) between information entities of the stream. Conversational pattern (stringent and low delay)	Voice (VoIP)
Streaming real-time	Preserve time relation (variation) between information entities of the stream	Streaming video
Interactive best-effort	- Request response pattern - Preserve payload content	Web browsing
Background best-effort	- Destination is not expecting the data within a certain time - Preserve payload content	Background downloads, non realtime

Figure 11.12 UMTS traffic classes in TS 23.107

QCI	Resource type	Priority	Packet delay budget (ms)	Packet error loss rate	Services	UMTS class (23.107)
1	GBR	2	100	1e-2	Conversational voice (VoIP)	Conversational
2		4	150	1e-3	Conversational video (live streaming, video call)	
3		3	50	1e-3	Real time gaming	Streaming
4		5	300	1e-6	Nonconversational video (buffered streaming)	
5	Non-GBR	1	100	1e-6	IMS signaling	Interactive
6		6	300	1e-6	Video (buffered streaming) TCP based (www, email, ftp, p2p filesharing, etc.)	
7		7	100	1e-3	Voice, video (live streaming), interactive gaming	Background
8		8	300	le-6	Video (buffered streaming) TCP based (www, email, ftp, p2p filesharing, etc.)	
9		9				

Figure 11.13 LTE traffic classes in TS 23.203 [4]

achieve quality for the service while offering a good business model. The macro model very much adapts to GBR, while the micro model is non-GBR. The GBR model can be translated to the expedited-forwarding (EF) class in the IETF DiffServ model, using high priority and weight and buffer tuning to achieve absolute service.

Returning to the LTE QOS and the E-UTRAN and EPC, one question is whether four to nine classes in radio QOS can be mapped to the classic packet-based well-known classes following the IETF DiffServ model. If we step back from academic solutions and demands to a practical approach, we see that LTE has four classes and two or three priority levels, a number that is very similar to the well-known QOS classes in today's packet-based networks: network-control (NC), EF, assured-forwarding (AF), and BE.

Control and signaling, and voice are two classes (NC and EF) that can be grouped into the same priority. These two traffic types cannot be over-provisioned or dropped. Instead, they require guaranteed delivery to achieve the desired quality level. Traffic volume estimates for voice on LTE networks are similar VoIP and SIP, as discussed in Chapter 4. For example, if each SIP call requires 80 kbps in one direction, the estimated number of VoIP sessions needs to fall within these bandwidth and delay demands.

The bulk of traffic, that is, the TCP traffic, is in the BE class. Surfing the web, email, chat, and so forth are all classic BE applications. With mobile networks, handheld TVs and video clips are two interactive services that are driving the increased bandwidth demands and that are used to justify GGSN and LTE capabilities. However, streamed compressed media content in unicast format and long-lasting sessions are a perfect fit for the BE class of service because neither is a true real-time broadcasting service. Instead, they are long-lasting TCP sessions used to download and watch compressed video content. Neither is very different from ordinary web service usage, and they can be managed well with TCP mechanisms. In practice, there are no differences between buffered and background TCP content if we speak in terms of QCI classes.

One of the real challenges is handling true broadcasting in the form of real-time, multicast streaming media. An example of one such application is video conferencing. This traffic follows the same structure as streamed media, as discussed in Chapter 4. It is loss sensitive, but to some extent can handle delay that results from local playback and buffering. QOS for online broadcasting is probably one of the biggest challenges for mobile solutions.

Figure 11.14 shows an example of mapping traffic classes in an LTE/UTRAN between radio and packet QOS.

IETF class	3GPP class	Priority	Weight (%)	Buffer (%)	Traffic types
Network-control (NC)	5/ Conversational	High	10	10	Network control traffic
Expedited-forward (EF)	1/ Conversational	Strict high	25	25 (alt x ms of traffic)	Voice and voice signaling
Assured-forwarding (AF)	2/3/7 Streaming	Medium/low	50	50	Streaming bigger packets
Best-effort (BE)	All others	Low	25	25	TCP and buffered data

Figure 11.14   Example of mapping traffic classes between the IETF and 3GPP

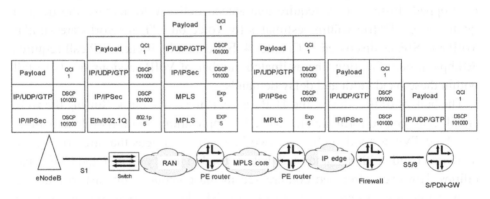

Figure 11.15   Example of traffic class mapping IETF<->3GPP

Let us follow a packet between eNodeB and S-GW/PDN-GW. Figure 11.15 shows the QOS PHB changes to this packet. Many media and headers can be involved, including DSCP, 802.1p, and MPLS EXP.

11.9   Conclusion

Packet network designers are concerned with the ratio of over-provisioning, while radio network designers focus on quality of service. IP RAN and QOS design needs to fulfill both of these world views at the same time. The micro and macro angles are very much in the spotlight here regarding how to achieve quality for the service while supporting a good business model.

QOS in an LTE RAN and backhaul network depends on QOS policies such as marking and queuing, as well as handling resilience and fault recovery. The following points should always be considered for QOS with IP RAN and mobile backhaul:

- Protection and fault recovery, and Ethernet OAM
- Synchronization
- Traffic separation and services

Protection and fault recovery of paths and links are essential for QOS services. Having an IP RAN that is dependent on a spanning tree is simply not good enough. In the case of Ethernet media, Ethernet OAM is the protocol that mimics the mature resilience of SDH and optical networks. While Ethernet OAM adds more features to the toolbox compared to plain Ethernet, which from the start was designed to be cheap and to allow easy provisioning of access media, it is not a good solution for transporting sensitive real-time applications. If the access between the NodeB and S-GW is a simple path provided by a carrier, the design needs to be adapted accordingly. If you have no control of the Layer 2 media because you rent an E-Line, VLAN, or something similar, more demands are placed on the end-to-end next layer, which is IP. If the eNodeB has an IP stack, why handle it like Layer 2? Layer 3-based OAM protocols are probably better suited to provide secure fault protection by triggering rerouting to an alternative path. An example is using BFD in conjunction with routing, either static or dynamic, to achieve dual-homed access for the eNodeB.

Clock synchronization design always comes down to using the design of choice. You can have an all-packet-based design such as RFC 1588 and NTP. Another option is to involve the media by using synchronous Ethernet. In this case, handling the clock handle requires some planning. Obviously, synchronous Ethernet is useful only if you have some control of the end-to-end Layer 2 paths. Otherwise, protocols that are more packet based, such as NTP, are more suitable.

As with most things in the telecom business, the most practical advice for the QOS policy design is "less is more." A good start is to separate the protocols and services into the three or four classes described earlier. Choosing these classes addresses the demands of both the architecture and the design of the IP RAN and mobile backhaul transport elements. If two levels of priority are enough, separating classes with the same weight can work. However, the cost and demands need to be taken into account. If you have control of the IP RAN and backhaul, a Layer 2 design could be sufficient and cost-efficient. Another consideration is network scaling. The same lessons from Internet growth need to be applied: Layer 3-based networks tend to scale well, while Layer 2-based networks do not. And the need for a Layer 3 network can happen faster than planned. If you do not control the paths, but instead rent an E-Line service, the demands and contracts need to be tight to secure sufficient uptime. Layer 3 processing is probably the only way to achieve good link management. The result of the cost savings of not having your own managed transport core is that your access equipment gets flooded with demands to handle protocols and management on layers above Layer 2 to compensate for the lack of end-to-end equipment control, for example, by adding routing and high-availability capability on the eNodeB access point to secure good availability on the path towards the S-GW and PDN-GW.

Earlier, we mentioned the protocol overhead for LTE networks because many protocols are involved for the traffic traveling between the NodeB and the S-GW. Several encapsulations can affect the eNodeB performance and QOS policy. Figure 11.16 illustrates the rewriting of code points into Ethernet and IP headers for a packet with Ethernet, 802.1q, IPsec, and GTP headers, resulting in multiple DSCP fields. The rewriting can result in many instructions to be performed if the forwarding capacity uses a central CPU-based arbitration on the eNodeB, or even on the S-GW.

The IP RAN and mobile backhaul network QOS design in a DiffServ model, that is, the model we describe in this book, is not that different from QOS in any other packet-based network. The scheduling and bandwidth ratio need to be adapted to the expected traffic volumes and to the business model. The QOS policy needs to be adapted to the traffic types. For example, real-time traffic cannot be over-provisioned or exposed to delay or packet loss to maintain quality.

At the same time, end user demands on services need to be in line with user contracts. If a user pays a fixed fee just to have Internet access and runs SIP as

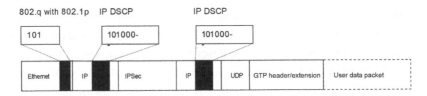

Figure 11.16   Classification and rewriting on the eNodeB

an application on their mobile handheld device, should the SIP traffic get better treatment than web surfing on that handheld? LTE networks will probably introduce differentiation on service levels even for real-time traffic based on subscriber policies.

To manage the policies and QOS profiles in a more dynamic way, a function called Policy and Charging Rules Function (PCRF) has been discussed for LTE networks. PCRF will work with MME to assist bearers in obtaining their QOS requirements and will be enforced on the S/PDN-GW. The idea is that when bearers are established, the QOS requirement from that bearer will be identified and negotiated with the PCRF. Then proper dynamic QOS profiles will be installed on the devices with which the PCRF communicates. PCRF is, however, still a work in progress.

References

[1] Wasserman, M. (2002) RFC 3314, Recommendations for IPv6 in Third Generation Partnership Project (3GPP) Standards, September 2002. https://tools.ietf.org/html/rfc3314 (accessed August 19, 2015).

[2] 3rd Generation Partnership Project (3GPP) (2009) 3GPP TS 33.210, 3G Security; Network Domain Security (NDS); IP Network Layer Security, Rel. 9.1.0, December 2009. http://www.3gpp.org/DynaReport/33210.htm (accessed August 19, 2015).

[3] Kent, S. (2005) RFC 4303, IP Encapsulating Security Payload (ESP), December 2005. https://www.ietf.org/rfc/rfc4303.txt (accessed August 19, 2015).

[4] 3rd Generation Partnership Project (3GPP) (2008) 3GPP TS 23.203, Policy and Charging Control Architecture, Rel. 10.0.0, March 2003. http://www.3gpp.org/DynaReport/23203.htm (accessed August 19, 2015).

[5] 3rd Generation Partnership Project (3GPP) (2005) 3GPP TS 23.207, End-to-End Quality of Service (QoS) Concept and Architecture, Rel. 9.0.0, September 2005. http://www.3gpp.org/DynaReport/23207.htm (accessed August 19, 2015).

Further Reading

3rd Generation Partnership Project (3GPP) (2004) 3GPP TS 33.310, Network Domain Security (NDS); Authentication Framework (AF), Rel. 10.0.0, September 2004. http://www.3gpp.org/DynaReport/33310.htm (accessed August 19, 2015).

Finnie, G. (2008) Policy, Identity and Security in Next Generation Mobile Networks, White Paper, Juniper Networks. www.juniper.net (accessed September 8, 2015).

Ghribi, B. and Logrippo, L. (2000) Understanding GPRS: The GSM Packet Radio Service. http://citeseerx.ist.psu.edu/viewdoc/summary?doi=10.1.1.20.7349 (accessed August 19, 2015).

12

Conclusion

A quality of service (QOS) deployment can be simple, maybe provocative but something the authors of this book strongly believe.

The key point is where and how to start. Understanding the foundations and the tools is critical, thinking about each tool as a black box that achieves one result. Analyze the traffic types present and who needs favoring and who gets penalized; there are no miracles or *win–win* situations. Also pay special attention to the network topology. Analyze the following: How is the traffic transported across the network? Is the transport protocol adaptive in any way? Can the network itself be lossless? And it is always equally important to keep it simple: use the features your deployment needs, and do not turn on any single new feature because it looks cool. Stay away from generic solutions. Think of it like a motorbike or car (depending if you are talking to Miguel or Peter); it is something tailored for your goals.

Everyone travels the same road, but ambulances can go first, while others wait for their turn. The tricky part of a QOS deployment is selecting what to classify as ambulances when the flashing lights show up.

In terms of future evolution there were times, which now feel like a very distant past, where it was argued that "just throw resources at the problem" approach placed a certain shadow over QOS. Not anymore, not just because overprovisioning is expensive but also because there are scenarios for which it is not tailored for.

First, we have the "real life" phenomena, traffic patterns are rarely flat and well behaved, and transient congestions such as microbursts are an unavoidable

QOS-Enabled Networks: Tools and Foundations, Second Edition. Miguel Barreiros and Peter Lundqvist.
© 2016 John Wiley & Sons, Ltd. Published 2016 by John Wiley & Sons, Ltd.

reality. Second, we have the predictability of the traffic flows, and this is more of a game changer. The location of traffic sources and destinations is no longer well known; for example, inside a data center a virtual machine could be talking with another virtual machine inside the same physical server, and then it is moved to another location. So traffic that was "hidden" inside a physical server is now crossing multiple network elements and links demanding resources. Multiply this phenomenon by the hundreds of virtual machines being moved around and the exact predictability of traffic flows becomes simply impossible.

In the first edition of the book in 2010, we wrote: "The role and presence of QOS will increase, and most QOS tools will become more refined and more complex as a result of industry input and requirements." Now look at any data center where a simple server has a rich set of QOS tools. So indeed QOS reach is spreading and tools are more and more refined, but there is no reason whatsoever for panic; if the reader understands the key concepts and tools, there is effectively nothing "brand-new" to learn.

Index

QOS-Enabled Networks: Tools and Foundations, Second Edition. Miguel Barreiros and Peter Lundqvist.
© 2016 John Wiley & Sons, Ltd. Published 2016 by John Wiley & Sons, Ltd.

Printed and bound by CPI Group (UK) Ltd, Croydon, CR0 4YY

27/10/2024

14580361-0001